Kinetic Theory

Student Monographs in Physics

Series Editor: Professor Douglas F Brewer
Professor of Experimental Physics, University of Sussex

Other books in the series:

Microcomputers
D G C Jones

Maxwell's Equations and their Applications
E G Thomas and A J Meadows

Oscillations and Waves
R Buckley

Fourier Transforms in Physics
D C Champeney

Kinetic Theory

J M Pendlebury

School of Mathematical and Physical Sciences, University of Sussex

Adam Hilger Ltd, Bristol and Boston

British Library Cataloguing in Publication Data

Pendlebury, J. M.
Kinetic theory.——(Student monographs in physics)
1. Gases, Kinetic theory of
I. Title II. Series
533′.7 QC175

ISBN 0-85274-796-9

Published by Adam Hilger Ltd
Techno House, Redcliffe Way, Bristol BS1 6NX, England
PO Box 230, Accord, MA 02018, USA

Printed in Great Britain by Page Bros (Norwich) Ltd

Contents

Preface

Kinetic theory has a great aesthetic appeal. Out of the chaotic motions of countless molecules come the simple and elegant laws of gas behaviour which have been confirmed by experimental observations to a most satisfying precision. In attempting to give such a concise account of the subject at a level suitable for undergraduate courses I hope that this attractive quality has been preserved.

I have tried to set out all the main assumptions in one place in Chapter 2 rather than holding some of them back for later stages in the development. The intention is to give the reader a clear view of what I feel is the minimum set of basic equipment needed to solve the simple everyday problems involving kinetic theory. Since I have included among them such things as the Maxwell–Boltzmann distribution of vector velocities and the principle of equipartition of energy, it does mean that these things have been introduced early and it may be that they will only acquire their full significance after reading on with frequent reference back to them. From Chapter 3 onwards I have made explicit use of currents, both of molecules per second and of momentum per second. For a small initial investment of effort in the ideas there is much benefit in unifying the approach to many of the arguments which follow.

With a subject of such mature years as kinetic theory the ground to be covered is well defined by common practice. Denied much flexibility in this respect, space has been at a premium and I have chosen to keep the text as full as possible and not to include sets of problems. As always there are other difficult decisions to make. I have tended to use the words *perfect gas* rather than ideal gas since this seems to have been the fashion for undergraduate level texts. I have been careful to say that the words perfect gas will imply a (real) gas for which the departures from the perfect gas laws are small enough to be neglected. This avoids the logical consequence of a perfect gas which obeys the perfect gas laws exactly having to have point molecules which, in principle, cannot rotate or vibrate internally. It is a pity that the clear distinction in the use of the words, with ideal gas for the former and perfect gas for the latter, as set out in Jeans' last book *The Kinetic Theory of Gases*, came too late to be adopted generally.

My enthusiasm for the subject has been enhanced over the years by many enjoyable discussions with Dr R Golub, Professor K F Smith and Dr W Steckelmacher. I am also most grateful to Professor D F Brewer, and Dr G K Woodgate for helpful comments on the manuscript.

Introduction

1

1.1 Objectives

A kinetic theory of gases seeks to explain the behaviour of gases in terms of the motion of their component molecules. The first step in setting up the theory is to adopt a set of basic assumptions about the nature of the molecules and how they move. Reasoning from them using the laws of mechanics, it should be possible to explain Boyle's law and all the other laws governing the behaviour of gases in equilibrium. It should also be possible to describe the behaviour of gases in nonequilibrium situations, such as where a gas is streaming through a small hole in a vessel into a vacuum space. Finally, it is of practical importance that the theory should be able to predict what will happen in new situations such as those, for example, which have arisen recently when very slow neutrons contained in vessels and transported along pipes have been found to behave in many ways like ordinary gases, but with some significant differences. A good theory should be able to achieve all these things without the need for further *ad hoc* assumptions. At the same time, the initial assumptions should be as few in number and as general in character as possible.

1.2 Brief Review of the Perfect Gas Laws

Frequent reference will be made to the gas laws which are obeyed by gases in equilibrium states where their pressures and temperatures are uniform throughout their volume. The particular laws listed below, which are known as the *perfect gas laws*, are an idealisation. Real gases at pressures of the order of one atmosphere do not obey them exactly, although their behaviour is very close to them and it gets closer and closer as the gas pressures and densities are made lower and lower.

(i) *Boyle's law*. As early as 1662 Robert Boyle, on the basis of experimental observations, stated that when a given mass of gas is compressed at a fixed temperature the product of its pressure and volume remains constant:

$$pV = \text{constant}. \tag{1.1}$$

(ii) *Charles' law*. In 1787 Charles, as a result of further experimental observations, found that when a given mass of gas is heated at constant pressure, the volume increases linearly with the temperature t on the Celsius scale according to the relation $V=V_0(1+\alpha t)$ where V_0 is the volume at 0 °C. He also found that the coefficient of expansion α is the same for all gases. Using the currently accepted value of α we may write

$$V=V_0\left(1+\frac{t}{273.15\,°\mathrm{C}}\right)=\left(\frac{V_0}{273.15\,°\mathrm{C}}\right)(273.15\,°\mathrm{C}+t). \tag{1.2}$$

Recognising the contents of the last bracket to be equal to the temperature, T, in kelvins on the thermodynamic (Kelvin) scale, we arrive at Charles' law in the form

$$V/T=V_0/273.15\,°\mathrm{C}=\text{constant} \tag{1.3}$$

which is illustrated in figure 1.1.

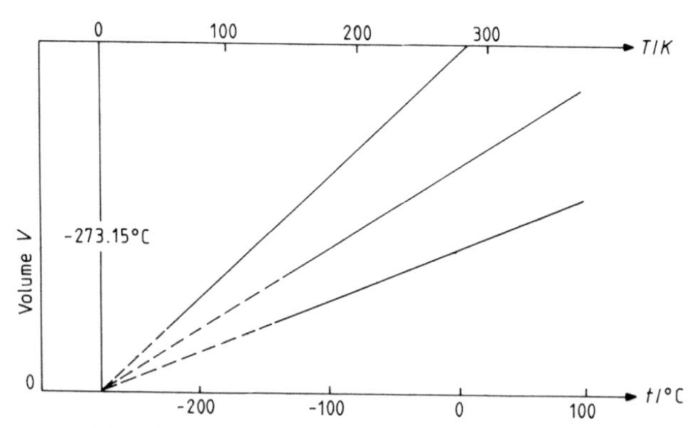

Figure 1.1 The relationship between volume and temperature for samples of three different gases.

(iii) *Equation of state for a gas*. Combining Boyle's and Charles' laws gives $pV/T=$constant. Measurements show that this last constant is proportional to the mass of gas in the sample. Furthermore, if we take one mole of gas the constant is found to be the same for all gases and is called the *gas constant*, R. Thus, for q moles of any gas

$$pV=qRT \tag{1.4}$$

which is the *equation of state for a perfect gas* or just the *perfect gas equation*. The best available value of R from measurements of pressures, volumes and temperatures, is $R=8.3144(3)\ \mathrm{J\,mol^{-1}\,K^{-1}}$. [Remember that one mole has a mass equal to the dimensionless relative molecular mass M_{M} of the substance times one gram, and it always contains the same number, called the Avogadro constant N_{A}, of molecules.]

(iv) *Joule's law.* From 1840 onwards Joule carried out a wide range of experiments to measure the energy content of various systems. By measuring the work and heat produced by gases when they expand, he found that the energy content of a gas is independent of its volume.

(v) *Dalton's law of partial pressures.* As part of his pioneering work on an atomic theory of matter *c* 1803, Dalton found that when a mixture of gases is present in a vessel, the total pressure on the walls is the sum of the pressures which the component gases would exert if each in turn was alone in the vessel.

(vi) *Gay-Lussac's law of combining volumes.* In 1808 the French chemist Gay-Lussac found that the volumes in which gases combine chemically and completely, when measured at the same temperature and pressure, are in a simple numerical ratio to each other, and also to the volume of the product if it is gaseous.

(vii) *Avogadro's hypothesis.* In 1811 Avogadro, an Italian lawyer who turned to physics, proposed that equal volumes of all gases measured at the same temperatures and pressures contain the same number of molecules.

1.3 Historical Perspective

Avogadro's hypothesis is the odd one out in the above list because in addition to rationalising experimental observation in terms of mathematical equations (laws) he had also adopted the idea that a gas is composed of molecules. In doing so he had introduced a theoretical *model* which, although it helped to explain several experimental observations, was speculative in that there could conceivably have been other ways of explaining the facts. He had taken Dalton's law of multiple proportions for weights of substances combining chemically, Dalton's hypothesis concerning atoms combining to form molecules, and Gay-Lussac's law of combining volumes, and by a process of careful reasoning, arrived at his hypothesis. By the same logic, he was also driven to conclude that gases like hydrogen and oxygen must have molecules which consist of two atoms tightly bound together. His ideas were not generally appreciated until their importance was demonstrated by the Italian chemist Cannizzaro in 1858. This was typical of the earliest phase of development of kinetic theory in which a number of isolated and speculative ideas were put forward by different people which for a long time gained little notice and even less acceptance. Much earlier, a Swiss mathematician and physicist Daniel Bernoulli had developed a theory involving the idea that the pressure of a gas on the walls of a vessel was due to a multitude of tiny impacts of component particles which were in constant rapid motion. This was included in his book on hydrodynamics published in 1738 which contained other major contributions to the theory of fluids for which he is even more famous. Despite this, few people knew of Bernoulli's kinetic theory, and in 1821 an Englishman, J Herapath, put forward a similar hypothesis about

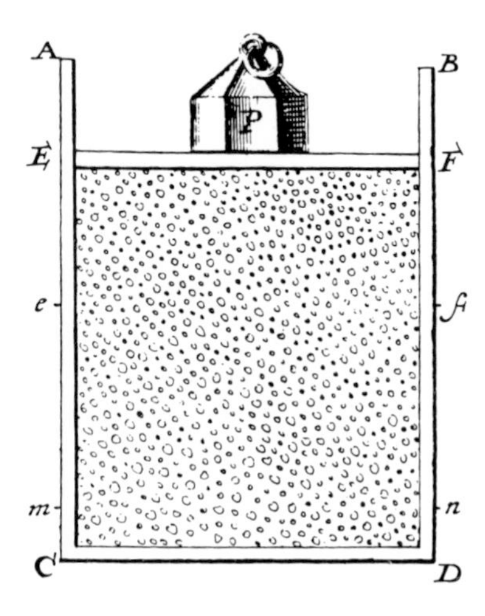

Figure 1.2 An illustration from D Bernoulli's book on hydrodynamics (1738), showing a weighted piston being supported by the particles of a gas.

the pressure of a gas. Other ideas were published by an English schoolmaster, J C Waterston, including that of the principle of equipartition of energy, in 1845. Prior to this time, the accepted theory of heat was that it was a fluid called caloric which could in some way percolate through all substances. It was not until the main battle to replace the caloric theory by the kinetic theory of heat was fought and won with the ideas and experiments of Count Rumford and Joule, that the other kinetic theory ideas had much hope of gaining general acceptance. Indeed, the work of Joule triggered off the first continuous and coherent phase of development of kinetic theory. This extended over the period 1850 to 1880 when the main foundations of the subject were laid by the work of Joule, Clausius, Maxwell, Boltzmann and van der Waals.

The following period, 1880 to 1930, was mainly one of consolidation and refinement in which a large section of theoretical techniques became known as statistical mechanics. Particularly notable contributions to the theoretical methods were made by the American J W Gibbs, the English physicist and mathematician Sir Sidney Chapman and the Swedish mathematician David Enskog. As a result of their work it became possible to obtain much more precise kinetic theory expressions for quantities such as viscosity, thermal conductivity and coefficient of thermal diffusion for a variety of laws of force between the molecules. This period also included further outstanding experimental work, notably that of Perrin, Knudsen, Estermann, and Stern. The work of Perrin (1910) on suspensions of fine solid particles in liquids was a particularly important landmark. It finally convinced the few remaining scientists who had remained sceptical of the kinetic theory, arguing that since there had been no direct observation of molecules, this theory might at some stage be replaced by another. The particles of fine powder which Perrin used could be seen in the microscope, although they were small enough to exhibit random thermal motion which Brown had observed with pollen grains in 1827. What Perrin was able to show with his careful quantitative measurements was that there was really no distinction, as far as the kinetic theory was concerned, between these solid particles and very large molecules. In that sense large molecules had finally been observed directly, and they were behaving exactly as the theory predicted.

During the latter part of this period the new quantum mechanics began to be incorporated in the kinetic theory. Happily, this actually simplified some of the calculations. It also helped in the understanding of physical phenomena such as the heat capacities of polyatomic gases which had previously been a great puzzle. However, these advances have not changed the underlying theme, which is to use the laws of mechanics applied to the behaviour of the component molecules to explain the behaviour of gases; it is just a matter of replacing classical mechanics by quantum mechanics where necessary.

Since 1930, developments in the theory have naturally been concerned with more complicated and unusual systems such as plasmas and diffusing neutrons, and in the calculation of effects which are sensitive to the details of the interatomic forces, such as thermal diffusion, surface adsorption, and the precise details of individual molecular scattering and reaction processes. On the experimental side, the use of molecular beams has become a feature of many important techniques such as those used in the vapour deposition of thin films, atomic spectroscopy, and devices such as the atomic clock which provides our time standard. Thermal diffusion in gases provided the first method for enriching uranium with respect to its isotope 235 and played a key role in initiating the production of nuclear weapons and nuclear power.

Assumptions of the Kinetic Theory

2

The early contributors to the kinetic theory, like Maxwell, had to think in a rather detailed way about all the collisions which the molecules make. In time, useful statistical generalisations emerged. We are now able to take advantage of these in formulating our set of basic assumptions. The assumptions are not all self evident. The reader may take it on trust that those of a statistical nature can be derived rigorously from mechanics by the methods of statistical mechanics, but this is outside the scope of this text. The assumptions also derive their justification from their great success in relating quantitatively a wide range of properties of gases, as the following chapters will try to demonstrate.

We shall need to make frequent use of the terms *thermal equilibrium*, by which we mean that the temperature is uniform throughout the system, and *mechanical equilibrium*, by which we mean that all the eddies and whirls generated in the gas when it entered the system have completely died away.

2.1 Nature of the Molecules

2.1.1 *Identical Molecules*

The molecules of a given type of gas are assumed to be identical, and in particular they will all have the same mass m. (In this respect different isotopes may be regarded as different gases.)

2.1.2 *Spherical Molecules*

For most kinetic theory purposes the molecules are assumed to be spherical. At first sight this may seem to be a rather crude assumption for anything other than a rare gas. The point is that even manifestly nonspherical molecules enter into collisions in all possible orientations. With this in mind it is then not so surprising that their collisional behaviour can be reproduced by a model which is some kind of spherical average over all orientations of the real molecule.

2.1.3 *Intermolecular Forces*

The typical behaviour of the mutual potential energy V between two molecules is shown as a function of the separation R between their centres by the full line of figure 2.1. The force between them is given by the relation $F_R = -\mathrm{d}V/\mathrm{d}R$. The force can be seen to be zero at what is called the equilibrium separation R_0. We can also conveniently introduce a radius R_1 beyond which the force is small enough to be neglected to the accuracy to which we are working. At separations less than R_0 the force is strong and repulsive, whilst at separations larger than R_0 it is the weak van der Waals force of attraction between transitory induced electric dipoles. The overall behaviour is often represented by the expression $V(R) = 4\,\Delta E[(\sigma/R)^{12} - (\sigma/R)^6]$ put forward by Lennard-Jones in 1931; this has been used for figure 2.1.

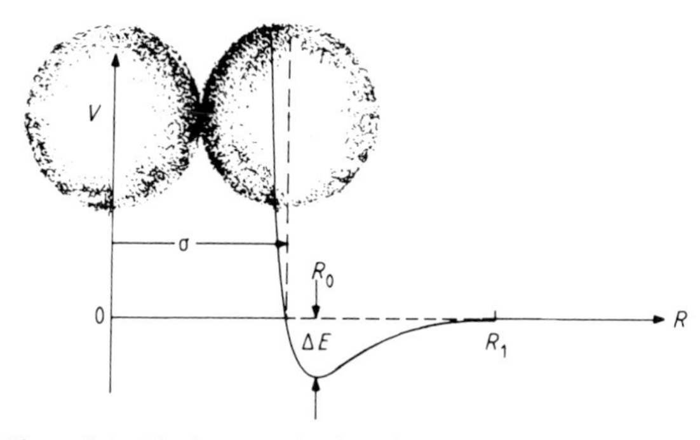

Figure 2.1 The heavy unbroken line shows how the mutual potential energy V of two molecules varies with the separation R between their centres.

For some purposes, such as the theory of viscosity, a simpler assumption about the forces is adopted which treats the molecules like hard billiard balls of diameter σ. Then the function V has the form shown by the broken line in figure 2.1. Values of σ for monatomic and diatomic molecules, found by comparing kinetic theory predictions for such things as the viscosity of gases with values from experiment, are mostly in the range $(2.5\text{–}4.5) \times 10^{-10}$ m. Values of ΔE are usually between 3×10^{-23} J (2×10^{-4} eV) and 8×10^{-21} J (5×10^{-2} eV). It is worth noting that, at the boiling point, the mean kinetic energy of the molecules is approximately $2\,\Delta E$.

In a solid or a liquid, the molecules always have nearest neighbours at a distance approximately equal to R_0 and there is always a considerable potential energy of interaction between them. In contrast, the molecules of air at STP have nearest neighbours at an average separation about $10R_0$, and most of the time they are travelling freely through space oblivious of their neighbours. This feature of the gaseous state distinguishes it from liquids and solids. It leads to the idea that the most *perfect* or *ideal* gas behaviour is approached when the average separation between the molecules becomes large compared with their diameter σ, which is the case at low densities. The fact that the gas laws of §1.2 have been found to represent the behaviour of gases at very low densities is why we call them the *perfect gas laws*. Dry air at STP deviates from the perfect gas law $pV=RT$ by less than 1 part in 10^3. *When we call a gas a perfect gas the implication is that we are treating its deviations from the perfect gas laws, and from the perfect gas equation* $pV=RT$ *in particular, as small enough to be neglected.*

In summary, we note that to explain the perfect gas laws, all we will have to assume about the intermolecular forces is that they have a limited range, and that the molecular diameters are small compared with the average distance to their nearest neighbours, which is true at low densities. The molecular speed distributions of Chapter 4 are even more independent of the forces; the molecules simply need enough separation to be a gas. To explain the phenomena in Chapter 5 such as viscosity, which involve intermolecular collisions, we need to introduce a well defined effective diameter σ and the billiard ball model is often used; the results are rather insensitive to other details of the forces and whether or not a region of attraction is included. On the other hand, to explain the equations of state of gases at higher densities as in Chapter 6, it is found to be essential to have forces with both repulsive and attractive regions, as in the Lennard-Jones model.

2.2 Continuous Motion of the Molecules

It is assumed that the molecules are in continuous random motion, and that the liveliness of this motion is determined by the temperature. At a fixed temperature, the mean speed of the molecules will be constant, as will all the other features of the distribution of speeds.

2.3 Uniform Distribution of the Molecules in Space

When a gas has reached mechanical and thermal equilibrium in a closed vessel, the number of molecules per unit volume, called the *number density*, n, is assumed to be the same in *all* parts of the vessel. This even applies to deep narrow fissures and pores in the wall, provided their dimensions are sufficient to contain molecules in free flight, and provided that such a feature does not pass right through the wall to constitute a leak. This assumption is only valid when the potential energy of the molecules can be taken to be constant throughout the volume.

2.4 Isotropy of the Velocities

In order to flow into a vessel, a gas must have some bias in the directions of the velocities of the molecules. We assume, however, that once the entrance is closed, the gas trapped inside invariably progresses towards mechanical equilibrium and the velocities of the molecules become isotropic in all parts of the volume. By this we mean that all directions of velocity occur with equal frequency. We can try to slow down the progress towards this equilibrium state by choosing a vessel with a regular shape, such as a sphere, or a cube, and by arranging conditions such that molecules bounce specularly (i.e. with their angle of incidence equal to their angle of reflection). However, in practice, there will always be some degree of irregularity of the walls which randomises the directions of velocities, and inevitably, the gas will eventually come to mechanical equilibrium.

It is useful to represent the instantaneous velocities of all the molecules in a sample volume by drawing their velocity vectors $\boldsymbol{c}$ in *velocity space* as shown in figure 2.2. The isotropy results in a three-dimensional figure which has spherical symmetry like a spiny sea urchin. It is sometimes even more useful to represent each vector by just a point at the head of each vector. Such a point has the coordinates (c_x, c_y, c_z). (To simplify the notation we will in future call these velocity components u, v, and w.) The sample of gas is then represented in velocity space by a cloud of points which has spherical symmetry about the origin (see, for example, figure 2.3).

The progress towards mechanical equilibrium is independent of whether or not there is progress towards thermal equilibrium (except insofar as we must assume that there is no convection taking place). In gases the two usually take place together, but very slow neutrons (called ultra-cold neutrons), which can be bottled like a gas, exhibit this progress to mechanical equilibrium in a time of a few seconds whilst remaining in complete thermal isolation from the walls.

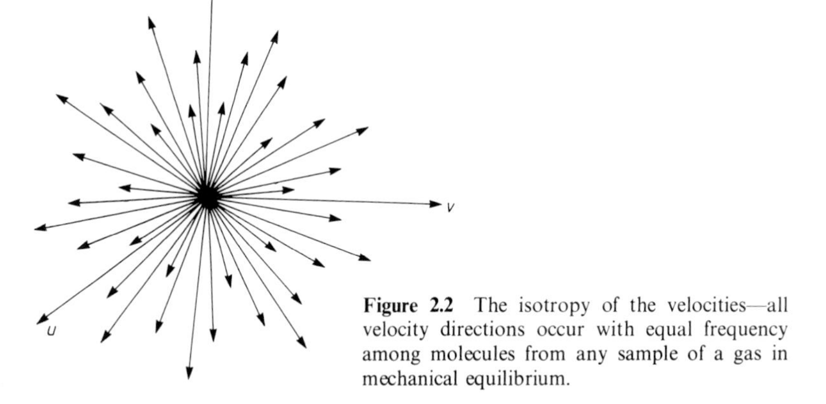

Figure 2.2 The isotropy of the velocities—all velocity directions occur with equal frequency among molecules from any sample of a gas in mechanical equilibrium.

2.5 Maxwell–Boltzmann Distribution of Velocities in Thermal Equilibrium

In the last section there were no assumptions about how the velocity points were distributed with respect to their speed c, which is with respect to their distance from the origin. If it is known that the gas has come to *thermal equilibrium* at temperature T with an *extensive medium* such as a vessel wall, we make the assumption that the number of molecules dn in unit volume of real space, with velocities in the small volume $du\,dv\,dw$ of velocity space where the velocity components are u, v, and w, is given by the Maxwell–Boltzmann velocity distribution:

$$dn = A\exp\left(-\frac{(\frac{1}{2}mc^2 + \text{PE})}{kT}\right)du\,dv\,dw \tag{2.1}$$

in which A depends on the gas density, k is called the Boltzmann constant, and PE is the potential energy of the molecules associated with their position in real space. The derivation of (2.1) requires the methods of statistical mechanics and is beyond the scope of this short text.

Figure 2.3 shows what the distribution of points in velocity space looks like for the common case where PE is constant. The density of points is greatest and nearly uniform close to the origin where c is small. It tails away smoothly towards zero at large speeds. This is related to the fact that large speeds imply large amounts of kinetic energy. However, total energy in a system is limited. The molecules may be regarded as competing amongst themselves, and with the molecules of the walls, for a share of the total energy in the system. Any molecule which becomes, as a result of chance collisions, very rich in energy, finds itself surrounded by hungry neighbours which are soon able to rob it of its excess energy. The general treatment of this sort of process was initiated by Boltzmann.

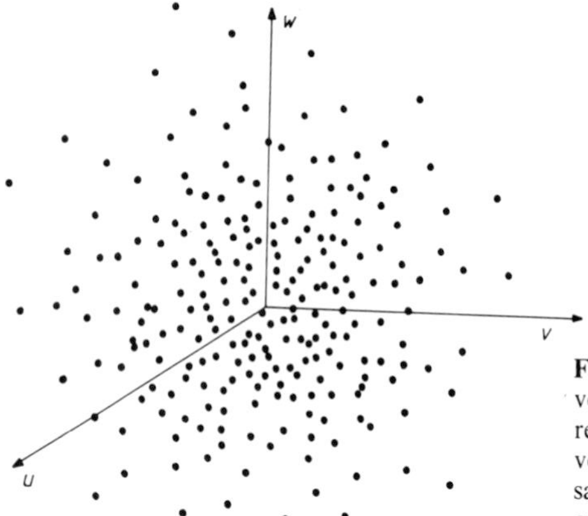

Figure 2.3 Cloud of points in velocity space. There is one point to represent the head of the velocity vector for each molecule from a sample of gas in thermal equilibrium.

The density of points in velocity space shown in figure 2.3 is given by (2.1) as $A\exp[-(\frac{1}{2}mc^2 + \text{PE})/kT]$. This exponential factor is known as the *Boltzmann factor*. It is useful to express it in the following ways:

$$\exp\left(-\frac{(\frac{1}{2}mc^2 + \text{PE})}{kT}\right) = \exp\left(-\frac{(\text{KE} + \text{PE})}{kT}\right) = \exp\left(-\frac{E}{kT}\right) = \exp\left(-\frac{3E}{2\varepsilon}\right) \tag{2.2}$$

where E is the total energy of the molecule, and ε is the average kinetic energy of a molecule. The next-to-last form in (2.2) is the most general one. The last expression shows how the Boltzmann factor may also be regarded as depending on how the total energy of a molecule compares with the average kinetic energy of the molecules. For vessels of normal size, the change of potential energy due to gravity is negligible compared with the kinetic energy of the molecules and we can set PE $=0$ everywhere, and the Boltzmann factor becomes $\exp(-mc^2/2kT)$, which is a gaussian-type function of c shown in figure 2.4.

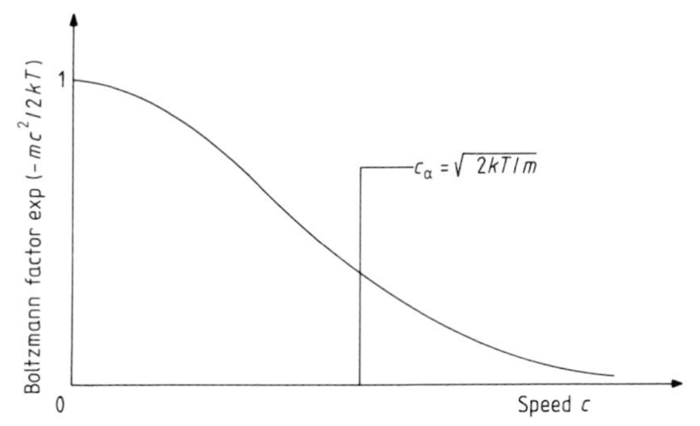

Figure 2.4 The Boltzmann factor as a function of the speed c of the molecules. The most probable speed c_α is defined in §4.2.

It should be noted that Boltzmann, working only from mechanics and using the laws of probability, could not immediately introduce a temperature T. He arrived at his factor in the form $\exp(-\beta E)$. To express the constant β in terms of temperature, it is necessary to apply the kinetic theory to the perfect gas law in the manner of §3.3.

2.6 Equipartition of Energy

First, we must introduce the concept of degrees of freedom. The total energy of a molecule will be made up of various contributions. The simplest case is that of a point molecule with zero potential energy for which $E=\frac{1}{2}mu^2+\frac{1}{2}mv^2+\frac{1}{2}mw^2$. In terms of independent Cartesian coordinates x, y and z for the position of the molecule, and using dots for time derivatives, this becomes $E=\frac{1}{2}m\dot{x}^2+\frac{1}{2}m\dot{y}^2+\frac{1}{2}m\dot{z}^2$. In general, a molecule has one *degree of freedom* for each of the squared terms involving the independent coordinates in the expression for E so that in this example there are three. The *principle of equipartition of energy* states that in *thermal equilibrium* at temperature T, the average energy per molecule, per active degree of freedom, is equal to $\frac{1}{2}kT$. The principle only applies under the following conditions:

(i) All forms of energy are in fact quantised so that only fixed discrete values are possible, and this applies even to the kinetic energy of translation. In ordinary sized vessels the allowed values of the latter are so closely spaced that it is hard to detect that they are not continuous. The condition for a degree of freedom to be *active* for equipartition is that the *spacing* of the energy levels associated with the type of motion which that degree of freedom involves should be much less than the quantity kT. This condition is easily satisfied for the translational motion in ordinary vessels at room temperature. However, it is not always satisfied, for example, in the case of the internal vibrations of diatomic molecules at room temperature.

(ii) If degrees of freedom associated with potential energy due to attractive forces between atoms and molecules are to be active, the forces have to be strong enough to produce bound (quantum) states of one molecule with another, with energies ranging over more than kT. The van der Waals forces are not strong enough to do this and they do not contribute active degrees of freedom in gases.

2.7 Detailed Balancing

This short text will not go quite far enough to use the principle of detailed balancing, but it is included here for completeness and because it is of intrinsic interest. The *principle of detailed balancing* states that for a system in *steady-state* equilibrium, the rate of transfer of molecules by collision from any particular state of motion A to any other particular state of motion B is equal to the rate of transfer by collision in the reverse direction from state B to state A. The principle is concerned with *direct* transfers by sudden processes which may be collisions with molecules, or with other particles, or with photons of radiation. As an example, the state of motion A might be defined as having a speed between c_1 and $c_1 + dc_1$, and state B as having a speed between c_2 and $c_2 + dc_2$. The word detailed is used to distinguish it from the more general and obvious steady-state condition where the number of molecules in any state A is constant. From the latter it follows that the total rate of transfer into the state A from all other states is equal to the total rate of transfer out of state A to all other states. The principle of detailed balance (rather than the general balance principle just stated) is illustrated in figure 2.5 for the example of the speed groups. The direct transfers which are taking place at equal rates are represented by the horizontal arrows.

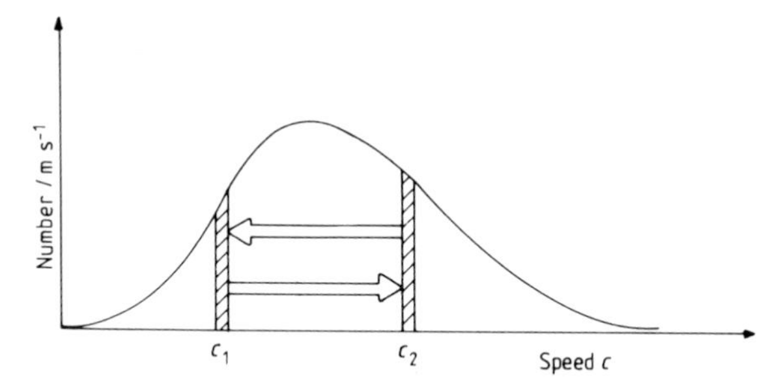

Figure 2.5 An example of detailed balancing. The arrows represent two equal rates of transfer between the speed ranges which are shown shaded.

Kinetic Theory of the Perfect Gas Laws

3

3.1 Pressure Due to the Impacts of Molecules

Consider a closed vessel containing a gas which is in thermal equilibrium at a fixed temperature, and in mechanical equilibrium with the walls. From our kinetic theory assumptions it follows that the distribution of molecular speeds and the average speed are unchanging, that the molecules are distributed uniformly over the volume, and that the distribution of velocities is isotropic throughout.

The basic idea is that the molecules of the gas are moving about rapidly in a random and very agitated way, frequently colliding with the walls and creating a pressure by their multitude of tiny impacts.

3.1.1 Molecule Partial Current

To begin with, we will calculate the rate of molecular impacts on an area A of the vessel wall shown shaded in figure 3.1. Let the area be small enough to be regarded as planar and choose axes with the z direction in line with the normal to A so that w is the velocity component directly towards the wall. In a small time interval $\mathrm{d}t$, a molecule will move a distance $w\,\mathrm{d}t$ towards the wall. We now fix our attention on that small volume shown in figure 3.1, from which molecules can arrive to hit the area A in the next small interval $\mathrm{d}t$. It has a base of area A and a

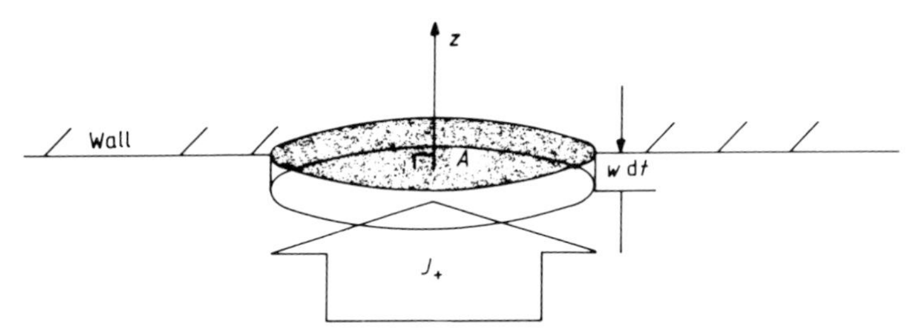

Figure 3.1 The small volume is that occupied by those molecules with a component of velocity w directed towards the wall which will collide with the area A in the next time interval $\mathrm{d}t$.

height $w\,dt$, so that its volume is $Aw\,dt$. At any instant this volume contains $n_w wA\,dt$ molecules with the velocity component w where n_w is the number density (number per unit volume) of these molecules. Thus, the rate of collisions on the area A is given by $n_w wA\,dt/dt = n_w wA$. If there are many different values of w called w_1, w_2, w_3, etc, with number densities $n_1, n_2, \ldots$, the total rate J_+ will be

$$J_+ = (n_1 w_1 + n_2 w_2 + \ldots)A. \tag{3.1}$$

Only positive values of w should be included in this summation, since molecules with negative values of w are moving away from the wall. We can conveniently make use of the average of the positive values of w to substitute for the above summation as follows. We will define the average $\overline{w_+}$ to be

$$\overline{w_+} = (n_1 w_1 + n_2 w_2 + \ldots)/n_+ \tag{3.2}$$

where n_+ is the total number density of molecules with positive w and the summation in the numerator is also over this set. (It is always important to be very specific in defining averages.) Combining (3.1) and (3.2) we find that

$$J_+ = n_+ \overline{w_+} A. \tag{3.3}$$

This is the *partial current*, in molecules per second, flowing directly towards and impinging on the surface A. The word partial has been used because it arises from just the molecules with positive values of w and is not the net current. It follows by symmetry from the isotropy of the velocities that $n_+ = n_- = \frac{1}{2}n$. In §5.1 we will show that $J_+ = \frac{1}{4}n\bar{c}A$, and hence that $\overline{w_+} = \bar{c}/2$. The factor of $\frac{1}{4}$ is seen to comprise one factor of $\frac{1}{2}$ because only one half the molecules is able to contribute, and a further $\frac{1}{2}$ arising from the average of the ratio w/c.

3.1.2 Momentum Partial Current

To calculate the force on the wall we need to know, not just how many molecules are flowing towards it per second, but how much momentum (in the direction of the normal) is flowing towards it per second. We must be careful to calculate the contribution from each value of w completely before making use of averages. Thus, for molecules with a particular value of w, there is a current of $n_w wA$ molecules per second flowing towards A, and each carries the component of momentum mw resulting in a momentum current of $n_w mw^2 A$. We see that the momentum current has mw^2, where the molecule current had just w. Proceeding as before and averaging over just those molecules taking momentum *towards* the wall, we may write the total momentum partial current $\dot{P}_+$ (momentum per unit time) as

$$\dot{P}_+ = n_+ m\overline{w_+^2} A. \tag{3.4}$$

A similar expression with the + signs replaced by − signs will describe the partial momentum current directly away from the wall due to the molecules with negative values of w.

From the fact that $u^2+v^2+w^2=c^2$, we see that on averaging over all the molecules, $\overline{u^2}+\overline{v^2}+\overline{w^2}=\overline{c^2}$. Since the velocities are distributed isotropically, their properties must not vary with direction, and we conclude that

$$\overline{u^2}=\overline{v^2}=\overline{w^2}=\tfrac{1}{3}\overline{c^2}. \tag{3.5}$$

It also follows by symmetry, from the isotropy of the velocities, that the distribution of values of w^2 will be the same for the positive w set as for the negative w set and hence:

$$\overline{w_+^2}=\overline{w_-^2}=\overline{w^2}=\tfrac{1}{3}\overline{c^2}. \tag{3.6}$$

Using (3.6) to substitute for $\overline{w_+^2}$ in (3.4) and replacing n_+ by $\frac{1}{2}n$ we find that

$$\dot{P}_+=\tfrac{1}{6}nm\overline{c^2}A. \tag{3.7}$$

The area A of wall absorbs a momentum current of $\frac{1}{6}mn\overline{c^2}A$ from $\dot{P}_+$ travelling towards it from the gas, and in addition suffers a further recoil due to creating a momentum current $\frac{1}{6}mn\overline{c^2}A$ for $\dot{P}_-$ travelling away from it as it projects molecules back into the gas. Whether the latter occurs by straightforward elastic reflection of the molecules, or by more complicated processes, does not affect the conclusion; the area A is responsible for a total rate of change of momentum of $\frac{1}{3}mn\overline{c^2}A$ which causes it to experience a force of this magnitude, directed along the normal away from the gas. It follows that the pressure, which is force per unit area, may be expressed as

$$p=\tfrac{1}{3}mn\overline{c^2}. \tag{3.8}$$

3.1.3 *On Hitting the Wall*

The molecules of a perfect gas are often assumed to bounce elastically and specularly from the wall, but the arguments used above show that this is not a necessary condition for obtaining the perfect gas equation. Real gas molecules are more likely to stick to the walls for a while, and then be ejected again later. For the gas to be in equilibrium the rate of sticking should be equal to the rate of ejection for each small element of wall area.

3.1.4 *Selecting Perfect Gas Behaviour*

In Chapter 6 we will start again with equation (3.8) and continue to develop the theory from it in a way which is applicable to gases at both high and low pressures. In the present chapter we will continue in a way which is restricted to perfect gas behaviour. It is important to say how this restriction is introduced. To avoid the need to consider molecules colliding with other molecules before hitting the wall we have worked in terms of the number density n extremely close to the wall and this is the meaning of n in (3.8). We will now assume that the surface n is the same as that in the bulk of the gas by substituting for n in (3.8) using the relation $n=N/V$ where N is the total number of molecules and V is the volume of the vessel. This is the step which restricts us to a perfect gas. In Chapter

6 we will show that there is generally a significant difference between the value of n close to the wall and that in the bulk of the gas due to the effects of intermolecular forces and collisions. However, it will also be apparent that the difference expressed as a proportion of n approaches zero as the density is decreased.

3.1.5 *Dalton's Law of Partial Pressures*

Suppose we have a gas in a vessel and that its pressure p is given by (3.8). In the next few sections it will be shown that the value of $\overline{c^2}$ for the molecules is just determined by the temperature T and the molecular mass m. Thus the value of $\overline{c^2}$ will not be changed if additional gases are introduced to the vessel at the same temperature. It follows from the remarks above that in the ideal gas approximation the number density of the first gas will also remain the same. Finally, since nothing has changed in (3.8) we can conclude that in the perfect gas approximation the partial pressure of a gas is unchanged when other gases are introduced into the same vessel.

3.2 Boyle's Law

Making use of the relation $n = N/V$ where N is the total number of molecules in the vessel of volume V, and substituting for n in (3.8) leads to the relation

$$pV = \tfrac{1}{3}mN\overline{c^2} \tag{3.9}$$

or

$$pV = \tfrac{1}{3}M\overline{c^2} \tag{3.10}$$

where M is the total mass of gas in the vessel. Whilst the temperature remains fixed, the value of $\overline{c^2}$ is constant, so that (3.10) is equivalent to Boyle's law.

3.3 Equation of State for a Perfect Gas

The perfect gas equation, based on the extrapolation of experimental measurements to low pressures, may be written for q moles of a gas as

$$pV = qRT. \tag{3.11}$$

The subject of mechanics does not provide a definition of temperature, so (3.9) is as close as any equation, based just on mechanics, can come to representing the perfect gas equation. *It is an assumption of the kinetic theory that the two equations* (3.9) *and* (3.11) *are equivalent*. One should note that this is the point at which the thermodynamic temperature T is brought into the kinetic theory, since according to this assumption we may write

$$pV=\tfrac{1}{3}mN\overline{c^2}=qRT=(N/N_A)RT \tag{3.12}$$

use having been made of the relation $q=N/N_A$, where N_A is the Avogadro constant.

3.4 Temperature and Molecular Kinetic Energy

By rearranging (3.12) we may write

$$\tfrac{1}{2}m\overline{c^2}=\tfrac{3}{2}(R/N_A)T. \tag{3.13}$$

This result, that the mean kinetic energy of translation of the molecules is proportional to T, imparts its own particular significance to temperature on the thermodynamic (Kelvin) scale. At this point, we have only established it for a perfect gas, but as indicated in the next two sections, it is true for gases in almost all conditions. The RHS of (3.13) also shows that the mean kinetic energy of the molecules does not depend on their mass, so that in two gases at the same temperature, one with heavy molecules and one with light molecules, both types will have the same average kinetic energy. If they are allowed to mix there will be no net transfer of heat between them. A simple mechanics calculation of the average energy transfer in collisions between molecules with different masses confirms that it is zero when their average kinetic energies are equal. Even a pendulum bob suspended in a gas will have a minimum kinetic energy of $3kT/2$ as it executes the equivalent of Brownian motion due to the impacts of molecules.

3.4.1 *Boltzmann's Constant*

Boltzmann's factor, which played a key part in his theory, was of the form $\exp(-\beta E)$ where β was an undetermined parameter. The application of his theory to the speeds of gas molecules, which will be covered in Chapter 4, shows that the mean kinetic energy of the molecules is $3/(2\beta)$, a result which is true for all gases. Identifying this with (3.13), in the case of a perfect gas, provides the key calibration which allows the constant β to be determined from experimentally measured quantities. Accordingly,

$$\tfrac{3}{2}(R/N_A)T=3/2\beta \tag{3.14}$$

whence

$$\beta^{-1}=(R/N_A)T=kT. \tag{3.15}$$

The constant k is called the Boltzmann constant and is defined as

$$k=R/N_A. \tag{3.16}$$

The current values, $R=8.3144(3)\ \mathrm{J\ mol^{-1}\ K^{-1}}$, and $N_A=6.02205(3)\times 10^{23}\ \mathrm{mol^{-1}}$, yield the result $k=1.38066(4)\times 10^{-23}\ \mathrm{J\ K^{-1}}$. Using this definition of k, the Boltzmann factor becomes $\exp(-E/kT)$.

3.4.2 *Equipartition of Energy*

The arguments of statistical mechanics also show that the average energy associated with each active degree of freedom is equal to $\frac{1}{2}\beta^{-1}$, which has just been shown to be equal to $\frac{1}{2}kT$. The principle of equipartition of energy is valid under the conditions set out in §2.6, and it applies to the kinetic energy of translation of gases under almost all practical conditions, not just in the perfect gas approximation. Given that the KE of translation involves 3 degrees of freedom, we can say that for all gases

$$\text{the average KE of translation per molecule} = \tfrac{3}{2}kT. \tag{3.17}$$

3.4.3 *Charles' Law*

In §3.3 we used Charles' law to introduce the temperature T into the kinetic theory via the perfect gas equation. It would not make sense, therefore, to claim that Charles' law provided any sort of test for the theory. Nevertheless, it is interesting to see that using it in this way has led to such a pleasing result as (3.17).

3.5 Avogadro's Hypothesis

Rearranging (3.12) we may write

$$pV = N(R/N_A)T = NkT \tag{3.18}$$

and

$$pV/T = Nk. \tag{3.19}$$

Avogadro's hypothesis states that equal volumes of all gases measured at the same temperature and pressure contain the same number of molecules. His gas samples, specified thus, all have the same value for pV/T. Equation (3.19) shows that they will, therefore, all have the same value of Nk and of N, in agreement with the hypothesis. The reasons which caused Avogadro to put forward his hypothesis were to do with proportions and weights, and did not involve the motion of the molecules. The fact that the same result has been obtained from the theory of the molecular motion is a significant success for the theory.

3.6 Calculating Molecular Number Densities and Speeds

3.6.1 *Number Densities*

Substituting $n = N/V$ into (3.18), which was $pV = NkT$, gives

$$p = nkT = n(R/N_A)T \tag{3.20}$$

and

$$n = p/kT. \tag{3.21}$$

Evidently, n is proportional to p at a given temperature. Noting that 1 atm = 101 325 Pa, 1 torr = 133.3 Pa = 1 mmHg and 1 Pa = 1 N m^{-2}, we find that for a perfect gas at 300 K

$$n/[\mathrm{m}^{-3}] = 2.45\times 10^{25} p/[\mathrm{atm}] = 3.22\times 10^{22} p/[\mathrm{torr}] \tag{3.22}$$

and also

$$n/[\mathrm{m}^{-3}] = 9.66\times 10^{24}\frac{p/[\mathrm{torr}]}{T/[\mathrm{K}]} = 7.24\times 10^{22}\frac{p/[\mathrm{N\,m^{-2}}]}{T/[\mathrm{K}]}. \tag{3.23}$$

In an ultrahigh vacuum at 10^{-10} torr, n is still as large as 3×10^{12} m^{-3}. In outer space n may fall to values as low as 10^4 m^{-3}.

The volume occupied by 1 mole at STP is called the *molar volume* V_A and may be calculated for a perfect gas using $V = RT/p$, with $T = 273.15$ K, and $p = 101\,325$ Pa, giving $V_A = 0.022\,413\,8(7)$ m^3.

3.6.2 *Molecular Speeds*

From the relation $\frac{1}{3}mN\overline{c^2} = (N/N_A)RT$ of (3.12) we may write

$$\overline{c^2} = 3kT/m = 3RT/mN_A = 3RT/M_A. \tag{3.24}$$

R will normally be expressed in J mol^{-1} K^{-1}, in which case M_A is the mass of one mole of the gas expressed in kg, which will be 10^{-3} kg times the relative molecular mass M_M (on the scale where the value for ^{12}C is exactly 12). Some examples of values of the RMS speeds obtained using the root of the last expression of (3.24) are shown in table 3.1.

Table 3.1 The RMS speeds/(m s^{-1}) of various molecules in thermal equilibrium at the temperatures shown.

Gas	M_M	77 K	300 K	1000 K
n	1	1.38×10^3	2.72×10^3	4.97×10^3
H_2	2	9.77×10^2	1.93×10^3	3.52×10^3
He	4	6.93×10^2	1.37×10^3	2.50×10^3
H_2O	18	3.27×10^2	6.44×10^2	1.18×10^3
N_2	28	2.62×10^2	5.17×10^2	9.44×10^2
Xe	131	1.21×10^2	2.39×10^2	4.36×10^2
I_2	254	8.70×10^1	1.72×10^2	3.13×10^2

Maxwell's Speed Distribution

4

Maxwell first derived the distribution for molecular speeds in 1859. His proof, and his conviction that the result was correct, were perhaps more a matter of intuitive genius than rigorous logic, and it is now accepted that the methods of statistical mechanics which grew out of Boltzmann's work provide a much sounder derivation. Maxwell realised that what he had discovered was very important because of its generality, and its usefulness. His distribution function applies to all gases, provided they are in thermal and mechanical equilibrium; it is not restricted to perfect gases, and it also applies approximately to neutrons in equilibrium with a surrounding medium. In this and succeeding chapters all the material can be taken to apply to gases in general unless specific restrictions are mentioned.

4.1 Derivation from the Distribution of Velocities

The distribution of molecular speeds can be obtained from the distribution of velocities which we assumed in §2.5. Figure 4.1 shows the geometry of the

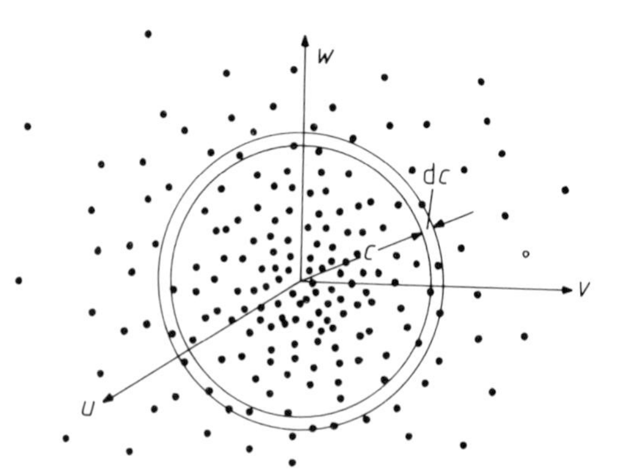

Figure 4.1 Finding the number of velocities which lie between two spheres corresponding to the speeds c and $c+\mathrm{d}c$.

situation in velocity space. As before, the dots represent the velocities of individual molecules in unit volume of the gas. The question is simple; how many of the velocities in the sample lie within the speed range c to $c+\mathrm{d}c$? It is the same as asking how many of the dots in the velocity space diagram lie in the thin spherical shell between spheres of radius c and $c+\mathrm{d}c$ centred on the origin. The volume of this shell is $4\pi c^2\,\mathrm{d}c$, and to get the number of points we simply multiply the volume by the number density of the points in the vicinity of the shell. (There is the same density of points in all parts of the shell because they are distributed with spherical symmetry about the origin.) Our kinetic theory assumption of §2.5 tells us that the appropriate density of points is $A\exp[-(\frac{1}{2}mc^2+\mathrm{PE})/kT]$ so that the number we want is given by

$$\mathrm{d}n = A\exp[-(\tfrac{1}{2}mc^2+\mathrm{PE})/kT]4\pi c^2\,\mathrm{d}c. \tag{4.1}$$

If we take our sample from a region where the potential energy of the molecules is constant, the factor $\exp(-\mathrm{PE}/kT)$ is also constant and can be absorbed into the A, as can the 4π. At the same time, it is convenient to take out a factor n which is the total number of molecules in the sample. These factors change A into a new constant B and we may write

$$\mathrm{d}n = nBc^2\exp(-mc^2/2kT)\,\mathrm{d}c \tag{4.2}$$

or

$$\mathrm{d}n = nf(c)\,\mathrm{d}c \tag{4.3}$$

where we are using the last equation to define the quantity $f(c)\,\mathrm{d}c$, which is the *probability* that a molecule, selected at random, will have its speed in the range c to $c+\mathrm{d}c$. The function $f(c)$ is called the *Maxwell probability distribution function for molecular speeds*, and it follows from (4.2) and (4.3) that it has the form

$$f(c) = Bc^2\exp(-mc^2/2kT). \tag{4.4}$$

The constant B can be found in terms of the other quantities by using the relation $\int_0^\infty f(c)\,\mathrm{d}c = 1$, which expresses the fact that each molecule is certain to have a speed somewhere in the range 0 to ∞. A few lines of calculus, using the methods described in the next section, give the result that $B=(4/\sqrt{\pi})(m/2kT)^{3/2}$. The function $f(c)$ is shown in figure 4.2.

4.2 Properties of the Maxwell Speed Distribution

The main features of this distribution are well worth remembering. Perhaps the first thing to note is that the distribution has a maximum at a particular speed which we call the *most probable speed* c_α, and clearly speeds in a broad band about this value are ones which are most likely to occur. Next, it is worth emphasising that the distribution remains high over a range of more than a factor of 2 in speed. Thus, any two molecules chosen at random are likely to have their speeds differing by as much as a factor of 2. There is, of course, a small

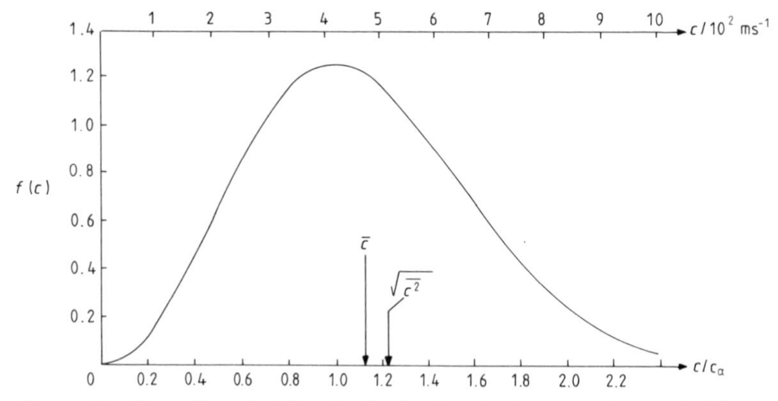

Figure 4.2 Maxwell's probability distribution function for molecular speeds. The top scale is for N_2 at 20 °C.

probability that their speeds will differ by a much larger factor. Thirdly, we see that, in spite of the fact that the density of points in velocity space peaks near the origin, the function $f(c)$ goes to zero at zero speed. If the ultracold neutrons mentioned earlier are to be bottled like a gas they must have speeds less than $c_\alpha/400$ when they come from a room temperature distribution. According to the Maxwell function, they have a number density n_u which is only 10^{-9} of the total n for all speeds. Experiments have confirmed that neutrons with these low speeds exist in the small numbers predicted. Going to the other extreme, we see that speeds which are many times the most probable speed are also very rare, due to the fact that the exponential factor becomes small very rapidly as c/c_α becomes greater than 2. The Maxwell distribution has also been compared with experiment (see figure 4.3) to a precision of about 2%, by looking at the speeds in a molecular beam, which are closely related to those in a gas, as will be shown in §5.2. The agreement is good except with high intensity beams where the slower molecules tend to get knocked out by faster ones coming up from behind.

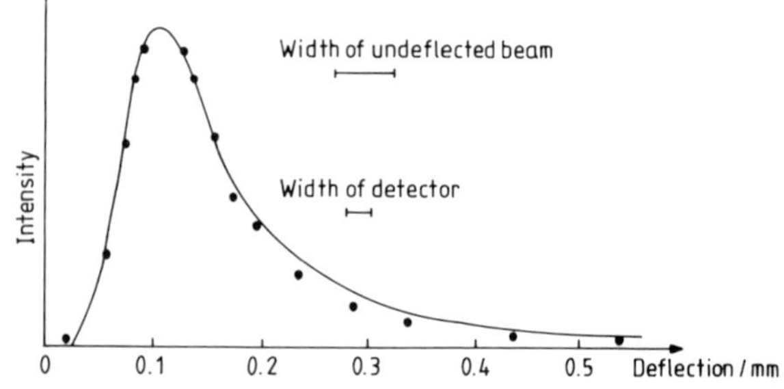

Figure 4.3 Vertical deflection of caesium atoms due to gravity after a horizontal flight path of about 1.5 m. Experimental points are compared with theory. (From Estermann, Simpson and Stern. *Phys. Rev.* **71** 1947.)

Turning now to the more mathematical properties of the distribution, it is convenient to think of it as having the form $Bc^2 \exp(-ac^2)$, where $a=m/2kT$. The following standard results are of use and can be found in most calculus texts:

$$\int_0^\infty c^{2n}\,\mathrm{e}^{-ac^2}\,\mathrm{d}c=\frac{1\cdot3\cdot5\ldots(2n-1)}{2^{n+1}a^n}\sqrt{\frac{\pi}{a}} \tag{4.5}$$

$$\int_0^\infty c^{2n+1}\,\mathrm{e}^{-ac^2}\,\mathrm{d}c=\frac{n!}{2a^{n+1}} \qquad (a>0) \tag{4.6}$$

where n is an integer. The $\int_0^\infty f(c)\,\mathrm{d}c$ needed for calculating B, as indicated in the previous section, can be found using (4.5) with $n=1$. It is also useful to calculate the mean values of the speed c, and of the square of the speed c^2. To do this we will make use of the general definition of a mean value. In the case of c, for example,

$$\bar{c}=\int_0^\infty cf(c)\,\mathrm{d}c=\frac{1}{n}\int_0^\infty c[nf(c)\,\mathrm{d}c]. \tag{4.7}$$

The second expression, although more clumsy, is seen to be the same as the first, and it can be interpreted as saying that in finding the mean, one sums the products of each value of the quantity and the number of times (square bracket) that it occurs in the sample, and then finally one divides by the total number n in the sample. To find the mean of c^2, the leading factor of c in (4.7) is replaced by c^2, and so on. The evaluation of $\bar{c}$ requires (4.6) with $n=1$, and that of $\overline{c^2}$ requires (4.5) with $n=2$. Remembering that c_α is the speed at which $f(c)$ has its maximum, which can be found by differentiating (4.4), the final results are

$$c_\alpha=\surd(2kT/m) \qquad \bar{c}=\surd(8kT/\pi m)=1.128c_\alpha \tag{4.8}$$

and

$$(\overline{c^2})^{1/2}=\surd(3kT/m)=1.225c_\alpha. \tag{4.9}$$

If it is more convenient, k/m can be replaced by R/M_A where R is the gas constant for a mole, and M_A is 10^{-3} kg times the relative molecular mass. Note that (4.9) is consistent with (3.13).

4.3 Distribution of One Component of Velocity

Returning to the Maxwell–Boltzmann velocity distribution, and making use of the fact that $u^2+v^2+w^2=c^2$, we can write equation (2.1), which involved $A\exp(-mc^2/2kT)\,\mathrm{d}u\,\mathrm{d}v\,\mathrm{d}w$, as

$$\mathrm{d}n=C\left[\exp\left(-\frac{mu^2}{2kT}\right)\mathrm{d}u\right]\left[\exp\left(-\frac{mv^2}{2kT}\right)\mathrm{d}v\right]\left[\exp\left(-\frac{mw^2}{2kT}\right)\mathrm{d}w\right] \tag{4.10}$$

where we have assumed that the factor involving the potential energy in (2.1) has been included in the constant C. The occurrence of u, v, and w in quite separate factors in (4.10) may be interpreted by saying that the frequency with which u lies between u and $u+\mathrm{d}u$ for given values of v and w is just the factor in square brackets involving u, which is quite independent of v and w. Thus the components of velocity are distributed independently of each other. This was taken as a key assumption by Maxwell in his derivation of the speed distribution, but it is difficult to justify without the methods of statistical mechanics which are needed to obtain (2.1).

If we wish to calculate the frequency with which u will lie between u and $u+\mathrm{d}u$, no matter what the values v and w are, we must integrate the RHS of (4.10) over the ranges $v=0$ to ∞, and $w=0$ to ∞. The results of the integrals can be absorbed into the normalising constant C, and taking out a factor of n for convenience, and calling the new constant B_u, we write

$$\mathrm{d}n = nB_u \exp\left(-\frac{mu^2}{2kT}\right)\mathrm{d}u = ng(u)\,\mathrm{d}u. \tag{4.11}$$

This defines the probability distribution function $g(u)$ for u, which is seen to be a gaussian function with a maximum at $u=0$, and which is symmetric with respect to positive and negative values of u. Using the same methods as in §4.2, B_u is found to be given by $B_u=\sqrt{(2m/\pi kT)}$.

As an interesting example, we note that $g(u)$ is the function needed to calculate the Doppler broadening of spectral lines in the light from glowing gas. If u is the velocity component of an emitting atom in the direction of observation, the received frequency ν is Doppler shifted relative to the original frequency ν_0 according to the relation $(\nu-\nu_0)/\nu_0=u/c_{\mathrm{L}}$ where c_{L} is the speed of light. Using this last equation, and the differential of it, to substitute for u and $\mathrm{d}u$ in terms of ν and $\mathrm{d}\nu$ in (4.11), we find that the number of molecules $\mathrm{d}n$ emitting frequencies between ν and $\nu+\mathrm{d}\nu$ derived from the original frequency ν_0, is such that

$$\mathrm{d}n \propto \exp\left[-\frac{mc_{\mathrm{L}}^2(\nu-\nu_0)^2}{2kT\nu_0^2}\right]\mathrm{d}\nu \tag{4.12}$$

which indicates that the light intensity will be a gaussian function of ν centred on ν_0. In the case of helium at 100 °C the full half width of the peaked distribution (4.12) is about $10^{-5}\nu_0$. (Note that $mc_{\mathrm{L}}^2/2kT$ is $c_{\mathrm{L}}^2/c_\alpha^2$.)

4.4 Effect of Potential Energy—the Isothermal Atmosphere

The atmosphere provides an example of a case where the potential energy of a molecule changes appreciably compared with kT over the height of the sample. It is instructive to see what happens to the speed distribution as we change height in an atmosphere, the whole of which is assumed to be in thermal equilibrium at temperature T. Returning to the Maxwell–Boltzmann distribution, (4.1), we can rearrange it as

$$dn = \left[(4\pi A/B) \exp\left(-\frac{\text{PE}}{kT} \right) \right] Bc^2 \exp\left(-\frac{mc^2}{2kT} \right) dc \tag{4.13}$$

which is of the form

$$dn = nBc^2 \exp\left(-\frac{mc^2}{2kT} \right) dc. \tag{4.14}$$

Evidently the shape of the distribution with respect to c does not change with height, a result which is reasonable since we took the temperature to be uniform throughout. Likewise, $\bar{c}$ will not change with height. There is, however, a change in number density. Identifying the number density n with the contents of the large square bracket of (4.13), and letting n_0 be the value of n where $\text{PE}=0$, we can deduce that

$$n = n_0 \exp(-\text{PE}/kT). \tag{4.15}$$

If the potential energy is due to gravity, $\text{PE}=mgh$, where h is the height, and

$$n = n_0 \exp(-mgh/kT). \tag{4.16}$$

The same result can be obtained by a simple mechanics argument: we consider a vertical column of the atmosphere, of cross section A, and fix our attention on a thin horizontal slice between the heights h and $h+dh$ as shown in figure 4.4(a). Equating the external upward force due to the pressure difference and the downward force of gravity on the slice, we have

$$-dp\,A = \rho A\,dh\,g \tag{4.17}$$

where ρ is the density of air at height h. Differentiating the equation $p=nkT$ with

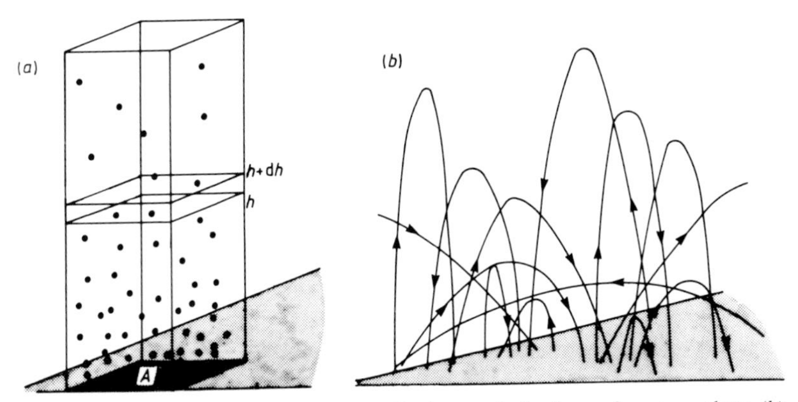

Figure 4.4 (a) Showing a thin horizontal slice in a vertical column of an atmosphere. (b) Trajectories of molecules forming a collisionless atmosphere in thermal equilibrium with the surface supporting it.

T constant gives $dp = dn\,kT$. Using this to substitute for dp in (4.17) and using the fact that $\rho = nm$ we arrive at the differential equation

$$dn/n = -mg\,dh/kT \tag{4.18}$$

for which the solution is the same as (4.16). We see that $n/n_0 = 1/e$ for $h = kT/mg = RT/M_A g$; for air at 0 °C this would be $h = 7900$ m (a little lower than the summit of Mt Everest).

Our original kinetic theory assumption of the Maxwell–Boltzmann distribution (2.1), from which all the results of this chapter have been derived, puts no conditions on whether or not the molecules are colliding with each other, nor does it rule out their being in a mixture of gases. This implies that (4.16) is also valid for an atmosphere where there are no collisions, in which we may picture the molecules being emitted from some horizontal surface at temperature T at the base of the atmosphere and then describing perfect parabolic trajectories until they fall down onto the surface again as shown in figure 4.2(*b*). Thus, in spite of the molecules slowing down under gravity as they rise up from the surface, the distribution of speeds must be the same and Maxwellian in shape at all heights! Again, it is only the number density which changes. As we move upwards, some of the slower molecules drop out, which helps to raise the average of the speeds and just compensates for the lowering effect caused by the slowing down of the others so that, over all, the average speed does not change.

4.4.1 *Perrin's Experiments on Sedimentation*

Equation (4.16) indicates that the heavier the molecules, the faster their number density decreases with height. There is nothing in its derivation which distinguishes between heavy molecules and small solid particles of matter. For one of his famous series of experiments, Perrin took the particles of various fine powders, suspended them in a liquid, and looked at how their number density varied with height after they had come to equilibrium. The mass m of each particle was so large that the number density varied appreciably over a height of a few tenths of a mm, and his observations in the microscope showed that the variations followed the exponential behaviour of (4.16). The weight mg was known from the diameter of the particles and the density of the substance of which they were composed. (He also had to correct for the buoyancy force due to the liquid.) From his measurements of the variation of n with height in a liquid of known density and temperature, he was able to deduce a value for the Boltzmann constant k, and, in turn, for the Avogadro constant N_A using the equation $N_A = R/k$. His value for N_A was in good agreement with that obtained by comparing the faraday of charge with the charge on the electron which had just been measured by Millikan. The observation that small but visible particles were behaving just like molecules was a spectacular success for the kinetic theory.

Collisions, Flow and Transport

5

5.1 Collisions with the Vessel Wall

5.1.1 Molecule and Momentum Partial Current Calculations

At this point we can complete the calculation of molecule partial current in a gas due to that half of the molecules which have a positive component of velocity in the direction of interest. The problem was introduced in §3.1 where we had shown in equation (3.3) that this current, in an arbitrarily chosen z direction, is given by the relation $J_+ = n_+ \overline{w_+} A$. We could now set out to calculate $\overline{w_+}$ but, unlike the case of $\overline{w_+^2}$, there is no powerful symmetry argument which we can use. Instead we will obtain J_+ by summing all the contributions from molecules with different w. Two things affect the size of w for any particular molecule: its speed c, and the angle θ between the velocity $\boldsymbol{c}$ and the z direction, via the relation $w = c \cos\theta$. We begin by restricting ourselves to molecules with speeds c which lie in the narrow range from c to $c + \mathrm{d}c$ of which there are $nf(c)\,\mathrm{d}c$, i.e. n times the probability of being in the range. The directions of their velocities are distributed *isotropically* over a total solid angle of 4π steradians, from which it follows that the number $\mathrm{d}n$ in any small solid angle $\mathrm{d}\Omega$ is given by

$$\mathrm{d}n = nf(c)\,\mathrm{d}c\,\mathrm{d}\Omega/4\pi. \tag{5.1}$$

If $\mathrm{d}\Omega$ involves directions which are all at an angle θ to the normal (see figure 5.1) all the $\mathrm{d}n$ molecules will have the same value of $w = c\cos\theta$ and the current $\mathrm{d}J_+$ $(= \mathrm{d}n\, wA)$ which they contribute may be expressed as

$$\mathrm{d}J_+ = ncf(c)\,\mathrm{d}c\cos\theta A\,\mathrm{d}\Omega/4\pi. \tag{5.2}$$

Let us now choose the element of solid angle $\mathrm{d}\Omega$ which selects *all* the directions with the same value of θ. The standard way of doing this is shown in figure 5.1, which concerns velocity space again. The group of all the directions between θ and $\theta + \mathrm{d}\theta$ to the z axis concerns the annular shaded area of the sphere. Its radius is $c\sin\theta$, its circumference is $2\pi c\sin\theta$, its width is $c\,\mathrm{d}\theta$, and its total area $\mathrm{d}S$ is $2\pi c\sin\theta\, c\mathrm{d}\theta$. By definition, the solid angle which it subtends at the centre of the sphere is $\mathrm{d}\Omega = \mathrm{d}S/c^2$, which in this case reduces to

$$\mathrm{d}\Omega = 2\pi\sin\theta\,\mathrm{d}\theta. \tag{5.3}$$

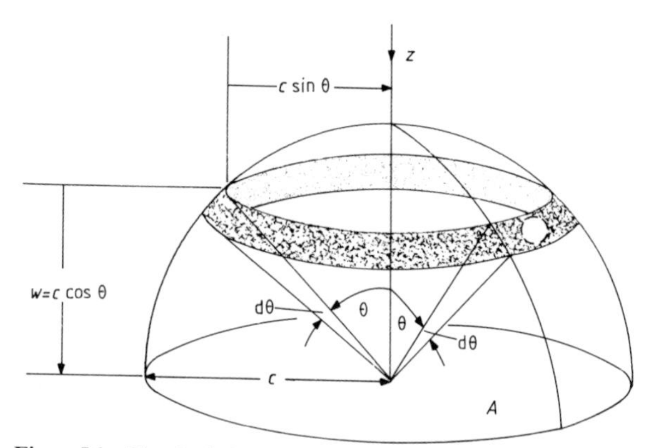

Figure 5.1 The shaded part of the sphere in velocity space accounts for all the velocities with speed c and with θ between θ and $\theta+\mathrm{d}\theta$.

Substituting for $\mathrm{d}\Omega$ in (5.2) using (5.3) gives

$$\mathrm{d}J_+ = ncf(c)\,\mathrm{d}c\tfrac{1}{2}\cos\theta\sin\theta\,\mathrm{d}\theta A. \tag{5.4}$$

To obtain the total current we first integrate over c from 0 to ∞ which yields

$$\mathrm{d}J_+ = n\bar{c}\tfrac{1}{2}\cos\theta\sin\theta\,\mathrm{d}\theta A \tag{5.5}$$

which we observe has a maximum at $\theta=45°$ indicating that the current is richest in molecules from directions close to 45° to the normal. Then we integrate over θ, from 0 to $\pi/2$ (which excludes negative values of w) and obtain

$$J_+ = \tfrac{1}{4}n\bar{c}A. \tag{5.6}$$

From (3.3), (5.6), and the fact that $n_+ = \frac{1}{2}n$, it follows that $\overline{w_+} = \frac{1}{2}\bar{c}$.

The calculation of the momentum partial current, which has already been completed by other methods in §3.2, can also be done using the methods above. For this purpose w is replaced by mw^2 just before (5.2), which causes $c\cos\theta$ to be replaced by $c^2\cos^2\theta$. As a result, $\bar{c}$ is replaced by $\overline{c^2}$ and the θ integral which is now over $\frac{1}{2}\cos^2\theta\sin\theta$, gives $\frac{1}{6}$ instead of $\frac{1}{4}$, leading once more to the result that the momentum partial current in the z direction is $\dot{P}_+ = \frac{1}{6}nm\overline{c^2}A$.

In §6.4 we shall discuss the existence of small differences between the values of n at the surface and in the bulk of the gas. Results (5.1) to (5.6) apply, regardless of whether the gas is perfect, provided the local value of n is used. However, at higher densities as described in §6.4, the above expression for the momentum partial current is only valid when applied to the gas very close to the surface of the vessel, and again the local n must be used.

5.1.2 *Cosine Laws for Mechanical Equilibrium*

Returning to (5.2) we find, after integrating over c, that

$$dJ_+ = n\bar{c}A\cos\theta\,d\Omega/4\pi. \quad (5.7)$$

Both (5.2) and (5.7) are expressions of what is called the *equilibrium cosine law* for molecules impinging on a surface when they come from directions within a solid angle $d\Omega$ at an angle θ to the normal. It applies to any $d\Omega$ satisfying these requirements—for example, either that subtended by the whole of the shaded annulus of figure 5.1, or one subtended by some part of it, like the empty circle.

In view of the isotropy of the velocities it follows that the currents dJ_+ and dJ_-, which exist within the same region, must be equal in magnitude, and have the same distribution with respect to the angle θ to the z axis. Hence,

$$dJ_- = n\bar{c}A\cos\theta\,d\Omega/4\pi \quad (5.8)$$

where, in this case, θ is the angle between the velocities and the normal pointing into the gas. Applying (5.8) to a region very close to the surface, it is clear that dJ_- is made up entirely of molecules which have just left the surface, so (5.8) is an *equilibrium cosine law* for molecules leaving the surface. It is quite a remarkable result, since it applies for the whole range of possible reflection laws which can operate when individual molecules arrive at the surface—from specular reflection at one extreme, where every molecule leaves with the same θ which it had when it arrived, through the cases where there is a partial correlation between leaving and arriving values of θ, to the other extreme of diffuse reflection where the direction of emission is quite independent of the direction of arrival. It should be emphasised that the word equilibrium used with these cosine laws means that there is a gas in mechanical equilibrium in front of the surface. Bearing this in mind, we conclude that *all* reflection laws for individual molecules have the property that integrating them over a cosine distribution for arrivals yields a cosine distribution for molecules leaving.

5.1.3 *Knudsen's Cosine Law*

It is a matter of logic to see that in the particular case where molecules leave a surface with no memory of how they arrived, the cosine law of emission must apply regardless of whether or not there is a gas in equilibrium in front of the surface determining the way in which molecules arrive. This is Knudsen's cosine law, which applies to the process of ejection after being stuck to a surface for a long time, and to evaporation. The *rate of ejection* from A (which is the current dJ_-) into a solid angle $d\Omega$ at θ to the normal, is proportional to $\cos\theta\,d\Omega$ (and hence into the range θ to $\theta + d\theta$ for all azimuthal angles round the normal it is proportional to $\cos\theta\sin\theta\,d\theta$). This will be true regardless of whether the surface is rough or smooth and regardless of any other of its properties—another important result and one which has been verified by experiment to about one part in 10^4 by Taylor in 1930.

5.1.4 *Mean Travel Distance Between Wall Collisions*

Consider a vessel containing n molecules per unit volume, all with speed c, in mechanical equilibrium. Imagine the molecules trailing threads as they go, like spiders. The total length of thread (i.e. of path) created in unit volume in unit time is nc, a quantity which is sometimes called the *total flux*. The total length of paths created in the whole vessel in time t is $ncVt$. The total number of wall collisions in this time is $\frac{1}{4}n\bar{c}A_v t$, where A_v is the total area of the vessel walls. Dividing the former by the latter gives the mean path $\bar{l}$ between wall collisions, first calculated by Clausius, to be

$$\bar{l}=4V/A_v. \tag{5.9}$$

The result (5.9) is not affected by intermolecular collisions, although when they occur they will cause the paths between the walls to have a zig-zag shape, and they also increase the statistical spread in the values of l.

5.2 Effusion and Molecular Beams

Suppose that a small aperture of area A_1 exists in the wall of an otherwise closed vessel containing a gas, and that the space beyond the aperture is pumped to a high vacuum. If the number density in the gas in the vessel is low enough for the *mean free path* (the mean distance of uninterrupted free flight) between collisions with other molecules to be much larger than the diameter of the aperture, the current of molecules J_+ arriving at the aperture will pass straight through, without making any collisions. After effusing in this way they continue to follow straight line paths, which fan out in the vacuum space as shown in figure 5.2. The distribution of molecules within J_+ can be taken from (5.2), which implies that, for the effusing molecules,

$$\mathrm{d}J=ncf(c)\,\mathrm{d}c A_1 \cos\theta \,\mathrm{d}\Omega/4\pi. \tag{5.10}$$

Thus, the molecular beams from A_1 obey the Knudsen cosine law, and the total current is $\frac{1}{4}n\bar{c}A_1$. The fractional loss rate of the contents of the source vessel is

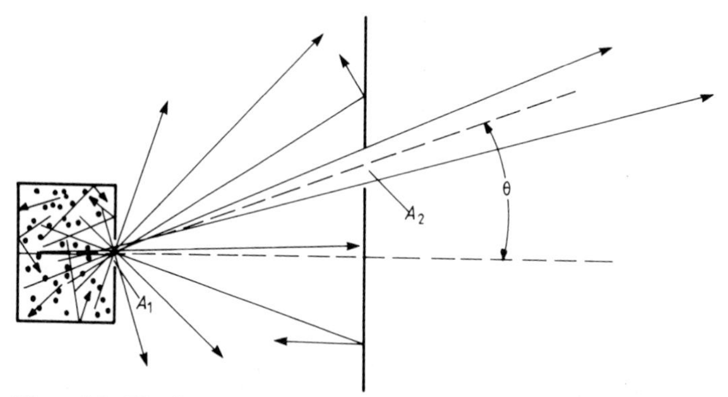

Figure 5.2 The formation of molecular beams. Molecules emerge from the vessel into a high vacuum through the aperture A_1. A wall with a second aperture A_2 can be used to form a collimated beam.

$\frac{1}{4}\bar{c}A_1/V$. This is proportional to $1/\sqrt{m}$ which is *Graham's law*. Beyond a second wall with an aperture A_2 there will be a more restricted range of velocity directions in what is called a *collimated* molecular beam. If the distance r between the centres of the apertures is large compared with their diameters the solid angle $d\Omega$ subtended by the second aperture at any point in the first is given to a good approximation by $d\Omega = A_2 \cos\theta/r^2$. Using this equation, and the Maxwell $f(c)$ given by (4.4) to substitute in (5.10), we find that for the beam current passing through A_2

$$dJ_B = \left(nB\cos^2\theta \frac{A_1A_2}{4\pi r^2}\right)c^3 \exp\left(-\frac{mc^2}{2kT}\right)dc. \tag{5.11}$$

It is worth noting that this distribution of beam current has the factor c^3 where the Maxwell speed distribution had c^2. (This is because the faster molecules have a greater probability of finding the aperture in a given time.) As a result the most probable speed c_{pB} and the mean speed $\bar{c}_B$ etc in the beam are all a little higher than the equivalent speeds in the gas. Using the standard integrals (4.5) and (4.6), one finds that

$$\int_0^\infty c^3\, e^{-mc^2/2kT}\, dc = \frac{1}{2}\left(\frac{2kT}{m}\right)^2 \qquad c_{pB} = \sqrt{\frac{3kT}{m}} = 1.22c_\alpha \tag{5.12}$$

$$\bar{c}_B = \frac{3}{4}\sqrt{\frac{2\pi kT}{m}} = 1.33c_\alpha \qquad (\overline{c^2})_B^{1/2} = \sqrt{\frac{4kT}{m}} = 1.41c_\alpha. \tag{5.13}$$

The speed distribution can be measured by placing a shutter in the beam, which is arranged to allow the molecules to pass in very short bursts. The distribution of speeds can be obtained from the distribution of *times of flight* for the molecules arriving at a detector which is some known distance from the shutter. An alternative method is to use deflections of a beam caused by sideways forces such as gravity, as described in figure 4.3.

5.3 Evaporation

The process of making thin films of metals and semiconductors by heating a lump of source material in a vacuum so that it evaporates, forming beams of molecules which collide with and stick to a solid substrate, has become a very important practical technique (see figure 5.3). An interesting question arises as to how the rate at which molecules leave the surface is related to the temperature of the source material. It is helpful to consider what would be happening if the source were in equilibrium with its saturated vapour at pressure p_T. An area A of the material would be accreting molecules from the vapour at the rate $K\frac{1}{4}n\bar{c}A$ where K is the *sticking probability* for a molecule striking the surface. The relevant number density is $n = kT/p_T$, and $\bar{c}$ is given by (4.8). For the vapour to be in equilibrium, molecules must be leaving the surface and returning to the vapour at the same total rate. Thus the last few simple equations can be used to

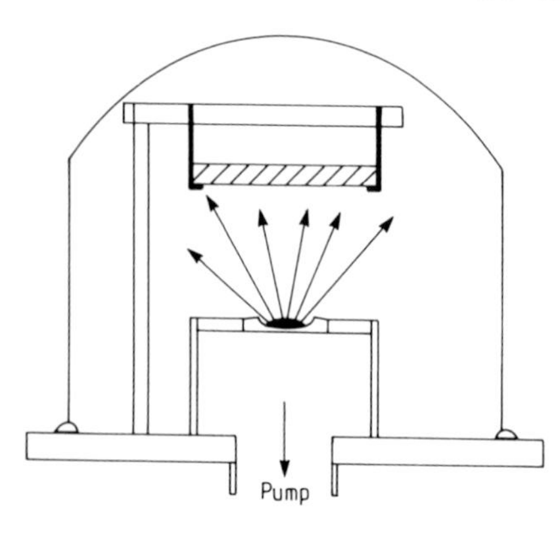

Figure 5.3 Forming a surface film by evaporating onto a substrate in a vacuum chamber.

calculate this rate provided we have values for p_T and for K. Knudsen carried out the first experiment to find a value for K, which is also called the *coefficient of evaporation* and which for any substance has a maximum value of unity. In the case of liquid mercury, he found that provided the surface was clean, K was close to unity, but that surface contamination could reduce K by orders of magnitude.

5.4 Thermal Molecular Pressure

Consider two chambers as shown in figure 5.4 which are maintained at different temperatures T_1 and T_2 and which are connected by a small aperture of area A. Suppose they contain a gas which is at low pressure so that the mean free path between intermolecular collisions is much greater than the diameter of the aperture. When the system has reached the steady state, the molecular currents passing in opposite directions through the aperture must be equal i.e. $\frac{1}{4}n_1\bar{c}_1A = \frac{1}{4}n_2\bar{c}_2A$. We have assumed that the molecules arriving at the aperture made their last collision deep inside the chamber which they are leaving and that their speed corresponds to the temperature of that chamber. Combining this last equation with the relations $p_1 = \frac{1}{3}mn\overline{c_1^2}$, and $\frac{1}{3}\overline{c_1^2} = \frac{1}{8}\pi\bar{c}_1^2 = kT_1/m$, etc, leads to the relation

$$p_1/p_2 = \sqrt{(T_1/T_2)}. \tag{5.14}$$

Similar pressure differences can be generated by temperature gradients along pipes, and it is something which one has to be aware of when making pressure measurements remotely via long pipes.

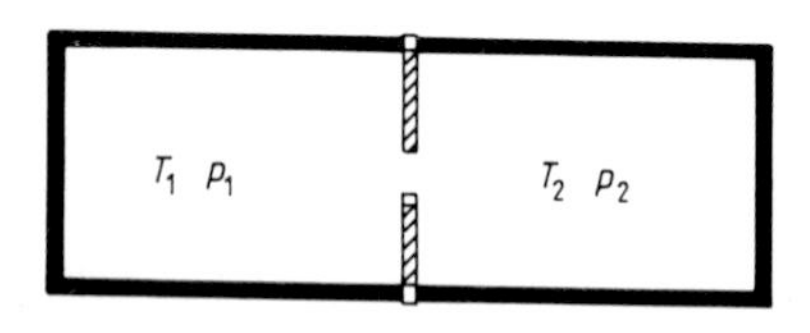

Figure 5.4 The pressures in two vessels maintained at different temperatures with an interconnecting aperture will not be the same.

5.5 Mean Free Path Between Intermolecular Collisions

The forces between molecules represented by the potential energy curve of figure 2.1 can significantly deflect their velocity directions if they are on course for their centres to pass within a distance of σ of each other. Such an encounter will be called a collision. Less close encounters will be regarded as producing deflections which are too small to matter. This amounts to treating the molecules as hard billiard balls of diameter σ.

5.5.1 Magnitude of the Mean Free Path

We begin by considering a molecule (labelled 1 in figure 5.5) which is travelling much faster than its neighbours, which can be treated as though they are at rest. The moving molecule can be thought of as sweeping out a cylindrical shaped volume with a radius σ and a length which reaches ct in time t. During the period t it will collide with any other molecules whose centres lie within this volume. Thus, the average number of collisions in travelling the distance ct is $n\pi\sigma^2 ct$, and the mean distance of free flight λ_{fast} travelled between collisions is

$$\lambda_{\text{fast}} = 1/n\pi\sigma^2. \qquad (5.15)$$

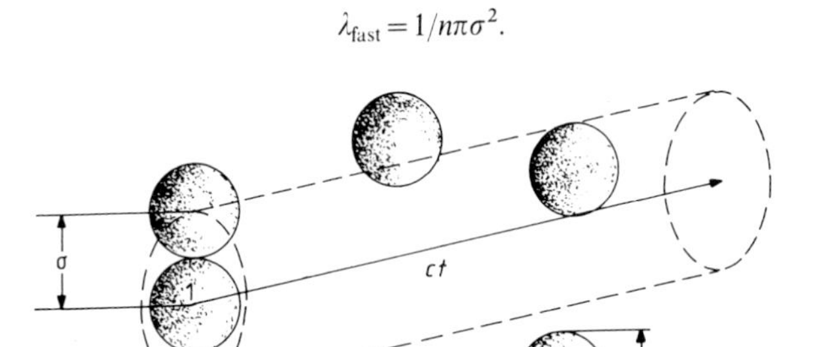

Figure 5.5 A fast moving molecule labelled 1 moving through a region with slow moving neighbours.

When the other molecules are travelling at the same speed as the one under consideration the calculation is more difficult. Clausius was able to show that the mean free path is then $\frac{3}{4}$ of that given by (5.15). The calculation becomes easy again in the other limit where the molecule of interest has a speed c' very much less than its neighbours. It can then be thought of as a sitting target which is being bombarded by the others. Their centres will bounce off a sphere of radius σ and surface area $4\pi\sigma^2$ centred on the molecule of interest. The average collision rate which it experiences is therefore $\frac{1}{4}n\bar{c}4\pi\sigma^2$. The reciprocal of this rate is the mean time between collisions and hence its mean free path between collisions is given by

$$\lambda_{\text{slow}} = c'/\bar{c}n\pi\sigma^2. \qquad (5.16)$$

Although we have not solved the problem completely, it has become clear that

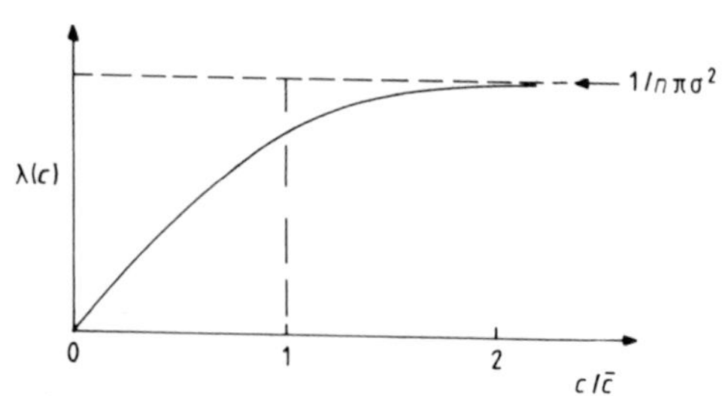

Figure 5.6 The way in which the mean free path of a molecule depends on its speed.

the mean free path has a speed dependence of the general form shown in figure 5.6. After a more detailed calculation in which the length of the free path is averaged over all cases assuming a Maxwell distribution of speeds for both molecules involved in a collision, the result obtained is

$$\lambda = 1/n\pi\sigma^2\sqrt{2}. \tag{5.17}$$

Without being more specific the phrase *mean free path* in a gas will generally refer to (5.17). (For very slow atoms or neutrons it is clearly important to use (5.16) instead.) Using the relation $n = p/kT$ to substitute for n in (5.17) we find that λ is inversely proportional to the pressure as expressed in the equation

$$\lambda = kT/p\pi\sigma^2\sqrt{2}. \tag{5.18}$$

Data on viscosity can be used to deduce that, for nitrogen at 20 °C:

$$\lambda/[\mathrm{m}] = \frac{4.8 \times 10^{-5}}{p/[\mathrm{torr}]}. \tag{5.19}$$

Table 5.1 shows the value of the equivalent multiplying constant for a number of other gases. If the pressures are expressed in pascals (N m^{-2}) the constants need to be multiplied by the factor 133.3.

There is a further small correction in which the λ given by these formulae should have approximately $\sigma/2$ subtracted to allow for the fact that molecules collide before the line of their centres is perpendicular to the relative velocity. Only at pressures above 1 atm is this a very significant change.

The mean time τ for which a molecule moves freely between collisions is given approximately by the relation $\tau \simeq \lambda/\bar{c}$. In air at 1 atm and 20 °C, τ is of the order of 100 ps.

Table 5.1 The number which, when multiplied by 10^{-5} and divided by the pressure in torr, gives the mean free path in metres for a variety of gases at 20 °C.

H_2	He	CH_4	Ne	CO_2	Cl_2	C_6H_6
9.0	14.2	3.9	10.1	3.2	2.2	12.1

5.5.2 *Distribution of Free Paths*

The distance which a molecule of a given speed will travel before making the next collision is subject to a large statistical variation from collision to collision. For a gas in equilibrium the environment of a molecule of a given speed is in a statistical sense unchanging with position and time, and we can say that the probability of making a collision in an interval of path $\mathrm{d}l$ is $\lambda^{-1}\,\mathrm{d}l$ where λ^{-1} is a constant which, as the notation anticipates, is the reciprocal of λ. Let the probability of a molecule *not* having made a collision, after travelling a distance l from the start of observations, be $p(l)$. Then $p(0)=1$ and $p(l)$ is getting smaller as l increases. The probability $-\mathrm{d}p(l)$ that a molecule makes its first collision (post $l=0$) between l and $l+\mathrm{d}l$ is equal to the probability $p(l)$ of reaching λ without a collision times the probability $\lambda^{-1}\,\mathrm{d}l$ of colliding between l and $l+\mathrm{d}l$, i.e.

$$\mathrm{d}p=p(l+\mathrm{d}l)-p(l)=-p(l)(\lambda^{-1})\,\mathrm{d}l. \tag{5.20}$$

The solution to this differential equation, with $p(0)=1$, is

$$p(l)=\exp(-l/\lambda). \tag{5.21}$$

The mean of the free path is given by the integral $\int_0^\infty l\exp(-l/\lambda)\,\mathrm{d}l/\lambda=\lambda$; which is consistent with the initial assumption about the relation between the constant and λ. The probability of free path l is shown as a function of l in figure 5.7, which also serves to show how a beam attenuates with distance due to collisions as it passes through a low density gas.

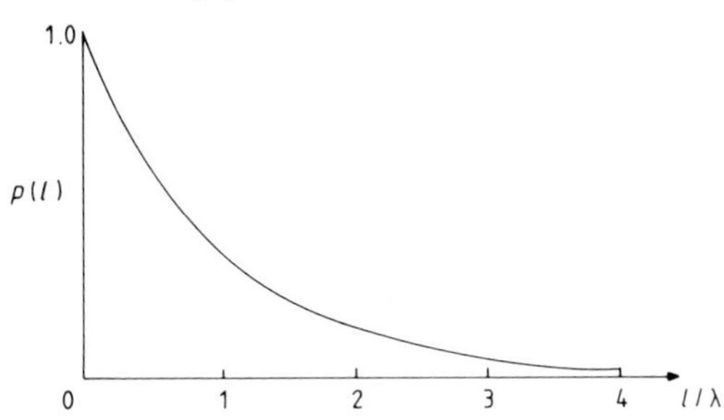

Figure 5.7 Probability distribution function for a collision-free path l against l/λ.

5.6 Transport Properties—Viscosity of a Gas

Situations involving the flow of gases down pipes are divided into two cases: (i) *viscous flow* where λ is less than the diameter of the pipe and (ii) *molecular flow* where λ is greater than the diameter of the pipe, which is usually the case when the pressure is below about 10^{-4} atm. We will begin by considering some properties in the viscous flow regime, starting with viscosity itself. Viscosity concerns the shear forces which exist when there is a gradient of the average drift (bulk flow) velocity across the gas. It is one of several properties which involve gradients in the gas where the gradient gives rise to a net transport, either of molecules themselves, or of some property which they are carrying, such as drift momentum in the case of viscosity, or energy in the case of thermal conductivity. If the gradients are gentle enough for the property not to change by more than a small fraction over a distance λ, we may employ some of the formulae obtained for gases in equilibrium.

In calculating the viscosity, let the *gradient* of drift (flow) speed s be in the z direction whilst $\boldsymbol{s}$ itself is in the x direction, as shown in figure 5.8. The definition of the coefficient of viscosity η allows us to say that the shear force F_s acting over an area A of the xy plane at position z is given by

$$F_s = \eta A \, \mathrm{d}s/\mathrm{d}z. \tag{5.22}$$

The kinetic theory explanation for viscosity is that there is a net current in the z direction through the area A of drift momentum in the direction of $\boldsymbol{s}$ which is responsible for and equal to the force F_s. We know that there is a current of molecules J_+ passing through the surface A in the positive z direction, and that

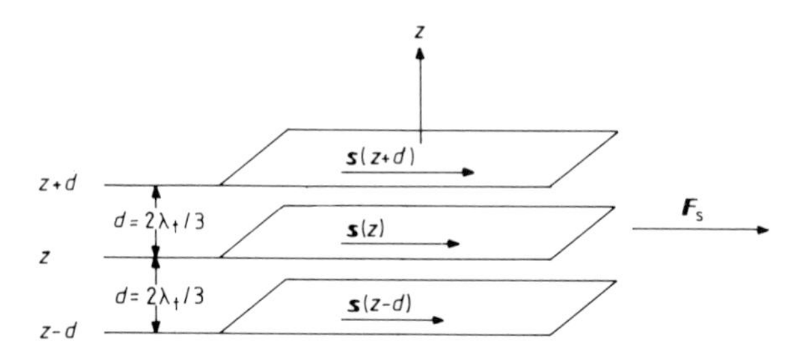

Figure 5.8

on the average they made their last collision in a plane at $z-d$, where d is a distance similar to λ. We assume that they are each carrying a drift momentum of $ms(z-d)$. Similarly, there is a current J_- of molecules travelling in the negative z direction, which have come from a plane at $z+d$ and which are assumed to be carrying a drift momentum $ms(z+d)$. Using the equations $J_+ = J_- = \frac{1}{4}n\bar{c}A$, and $s(z \pm d) = s(z) \pm (\mathrm{d}s/\mathrm{d}z)d$, we find that the net current of drift momentum in the $-z$ direction is $\frac{1}{2}n\bar{c}A(\mathrm{d}s/\mathrm{d}z)d$. After substituting this momentum current for F_s in (5.22) and using without proof the equation $d = 2\lambda_t/3$ where λ_t is called the *transport free path* we obtain

$$\eta = \tfrac{1}{3} nm\bar{c}\lambda_t. \tag{5.23}$$

To have obtained the numerical coefficient properly, we would have had to consider the contribution to the drift momentum current from each speed c and direction θ separately, performing the integrations at the last stage as we did when obtaining the formula $\frac{1}{4}n\bar{c}A$ in §5.1. There are other effects, such as the small persistence of the original velocity direction after each collision which Jeans estimated introduced a further factor 1.38, which is often allowed for by taking λ_t to be $\frac{4}{3}\lambda$. The more powerful theory of Chapman and Enskog yields $\eta = 0.499nm\bar{c}\lambda$ for hard sphere molecules. On substituting for λ using (5.18) this last expression for η becomes $\eta = 0.353m\bar{c}/(\pi\sigma^2)$ which shows that η is independent of density (and of pressure at a fixed temperature) and is proportional to $\sqrt{T}$. These properties are somewhat contrary to our normal expectations, which tend to be based on experience of viscosity in liquids.

If we are interested in the viscous drag between two parallel solid surfaces in relative motion with a separation L where $L < \lambda$ then we are concerned with the *molecular flow* regime. The molecular currents J_+ and J_- now pass between the two surfaces without making collisions on the way, each set of molecules having a drift speed characteristic of the surface which they have just left. Approaching the calculation as before, the only difference is that whereas in the viscous flow case the two sets of molecules travelling in opposite directions came from planes separated by $2d$ or $4\lambda_t/3$, they how come from planes which are separated by L. This replacement follows through to the result for *apparent viscosity*, $\eta_{ap} = \frac{1}{4}nm\bar{c}L$ which is seen to decrease with pressure. The function $\eta_{ap} = \frac{1}{3}nm\bar{c}\lambda_t/(1 + 4\lambda_t/3L)$ which reproduces the results in both regimes and makes a reasonable representation of the transition between them is shown in figure 5.9. This formula implies a small correction in the viscous flow regime, and indeed there is such a correction which relates to the fact that the variation of drift velocity with z is nonlinear at distances of less than one mean free path from the solid surfaces. At an infinitesimal distance away the molecules of J_- have exactly the same drift speed as the surface but those of J_+ have a speed characteristic of a plane which is $2\lambda_t/3$ away, so that between them they have an average speed which is different by a finite amount from that of the surface, a phenomenon which is called *surface slip*. The viscous drag in the molecular flow case may be reduced by a further factor if the molecules arriving at the surface do not give up all their drift momentum before leaving again.

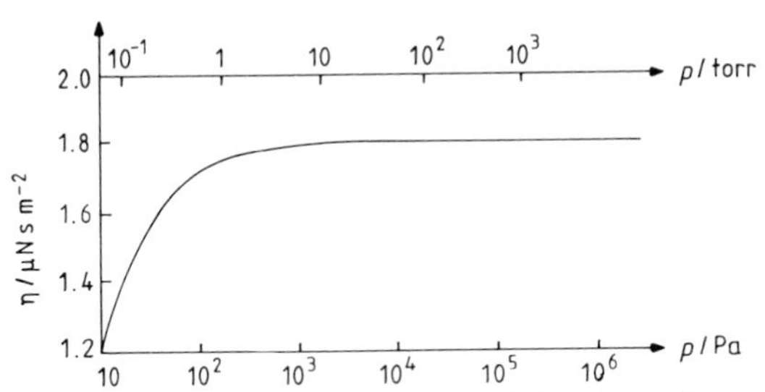

Figure 5.9 The apparent coefficient of viscosity against pressure for plane parallel surfaces separated by 1 mm of air.

5.7 Thermal Conductivity

In this case we are concerned with a gradient of temperature in the gas which again we will take to be in the z direction. The property being transported by the molecules is their contribution to the internal energy of the gas, which in the monatomic case would be almost entirely kinetic energy. If the internal energy per unit mass is $U(T)$ its value per molecule is $mU(T)$ and at the two planes of interest $mU(z \pm d) = mU(z) \pm m(\mathrm{d}U/\mathrm{d}T)(\mathrm{d}T/\mathrm{d}z)d$ so that the net current of internal energy, which is the rate of heat flow $\dot{Q}$, through the area A is given by

$$\dot{Q} = -\tfrac{1}{2}n\bar{c}Am\frac{\mathrm{d}U}{\mathrm{d}T}\frac{\mathrm{d}T}{\mathrm{d}z}d. \tag{5.24}$$

The numerical factor, which depends a little on the type of molecule, is larger here because of the high power of c which enters, and we will use $d = \tfrac{4}{3}\lambda_t$. The definition of thermal conductivity κ gives us the equation $\dot{Q} = -\kappa A\,\mathrm{d}T/\mathrm{d}z$. Using (5.24) to substitute for $\dot{Q}$, and the relation $\mathrm{d}U/\mathrm{d}T = C_V$ (per unit mass) we find

$$\kappa = \tfrac{2}{3}n\bar{c}mC_V\lambda_t = 2\eta C_V. \tag{5.25}$$

Most observations are consistent with (5.25) to within about 30%.

The differences which arise in the molecular flow case are similar to those we had with viscosity. For parallel plates at different temperatures separated by a distance L, the *apparent thermal conductivity* can be obtained by replacing $\lambda_t/3$ by $L/8$ in (5.25). There may be a further reduction in the effective conductivity because the molecules only transfer a fraction (called the *accommodation coefficient*) of their excess thermal energy to the surface before leaving again.

5.8 Diffusion of Molecules

A steady gradient of number density n causes a net current of molecules given by $J_{\mathrm{net}} = J_+ - J_-$. Taking the z direction in the direction of the gradient as before, we have the number densities $n(z \pm d) = n(z) \pm (\mathrm{d}n/\mathrm{d}z)d$ relating to the two currents $J_\pm$ so that $J_{\mathrm{net}} = -\tfrac{1}{2}\bar{c}Ad\,\mathrm{d}n/\mathrm{d}z$. Taking $d = \tfrac{2}{3}\lambda_t$ and using the usual definition of diffusion coefficient D to write the final expression, we have

$$J_{\mathrm{net}} = -\tfrac{1}{3}\bar{c}\lambda_t A\,\mathrm{d}n/\mathrm{d}z = -DA\,\mathrm{d}n/\mathrm{d}z \tag{5.26}$$

and hence

$$D = \tfrac{1}{3}\bar{c}\lambda_t. \tag{5.27}$$

In a normal gas under viscous flow conditions the pressure gradient associated with the number gradient causes bulk motion to take place in addition to the diffusion described by this last equation. The bulk flow, which is governed by the viscosity, is much larger than the diffusion current described above. However, there are situations of interest where the diffusing particles are a tiny minority dispersed among a uniform gas of different molecules. Any forces exerted on the host gas are too small to cause bulk motion. The net current of the minority molecules is then due to diffusion as described by (5.26) where n is their number density (excluding the host gas) and their mean free path relates to collisions with all kinds of molecules in the mixture. A relevant example is the spread of a trace of pungent gas released at one point in a room full of still air. Once the minority molecules are released from the source point they set out tracing zig-zag paths as each collision forces them to forget almost completely which way they were going before it happened. This kind of path is often referred to as a random walk; figure 5.10(*a*) shows some examples and indicates where the molecules have got to after a fixed time from a simultaneous release. Figure 5.10(*b*) shows the profile of the number density $n(x)$ along an axis x passing through the starting point, at different times after a group of particles is released simultaneously. Einstein showed that the mean distance of the particles from the starting point (as the crow flies) increases as the square root of the elapsed time. This distance is also very much smaller than the total length of a zig-zag path, which explains why it takes some seconds for a pungent gas to cross a room in spite of the very high speed of the molecules.

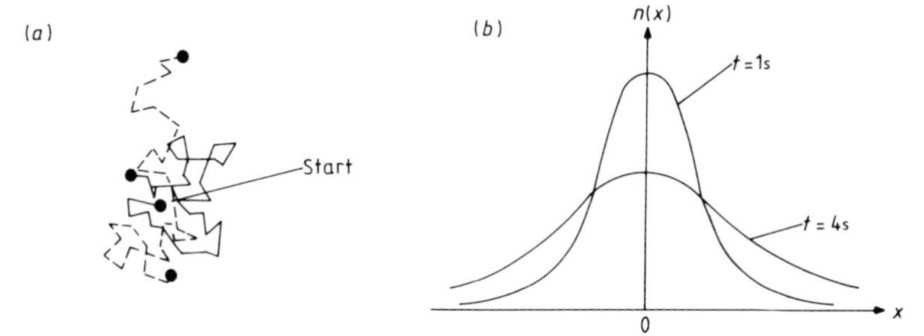

Figure 5.10 (*a*) Random walks for molecules released from the same point. (*b*) The number density along a line through the starting point after different elapsed times.

Equilibrium States of Gases in General

6

The aim of this chapter is to introduce simple kinetic theory explanations for the way in which the equilibrium states of gases in general relate to those of a perfect gas.

6.1 The Virial Equation

Given that the departures from the perfect gas law at a fixed temperature increase with density and therefore with pressure, it is useful to express the quantity pV as a series expansion in powers of p in the form

$$pV = qRT(1 + B'p + C'p^2 + \ldots). \tag{6.1}$$

This equation of state is called the *virial equation* and the coefficients B', C' ... which are of either sign and functions of temperature, are called the *virial coefficients*†. We already know that A' is 1. We shall pay most attention to the term which is linear in p involving the second virial coefficient B', since this is the term which dominates the first departures from ideal gas behaviour as the pressure is increased. For nitrogen the $B'p$ term is about -5×10^{-4} at 1 atm and 0 °C, whilst the $C'p^2$ term is 3×10^{-6}. The equation shows that the expression in parentheses totals pV/qRT which, for a perfect gas, is equal to 1. The departure of this quantity from 1 is a useful measure of the departure from ideal gas behaviour and we shall refer to it again.

The first precise experimental data for a gas over a very wide range of pressures and temperatures were provided by the work of Andrews on carbon dioxide over the period 1869 to 1876. Most real gases have the same general characteristics for the variation of pV with p which are shown schematically in figure 6.1 as a series of isotherms. (In the case of a perfect gas, this diagram would be a series of straight lines parallel to the p axis.) Gases differ widely in the absolute values of pressure and temperature at which the main features of these curves occur. As an example, each gas has a temperature, called the *Boyle temperature*, where the coefficient $B' = 0$ and the initial gradient of the isotherm is zero; some values are

† Unprimed constants A, B, etc are often reserved for the expansion of pV in powers of $1/V$.

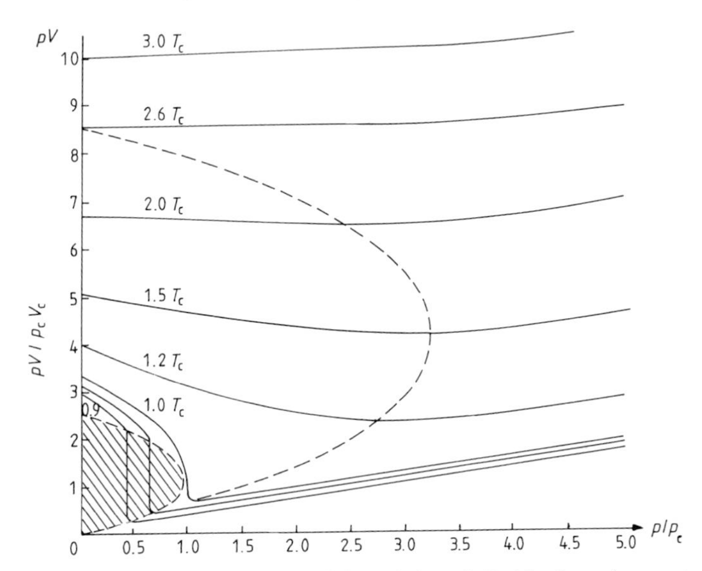

Figure 6.1 The general character of the variation of pV with p for real gases at various multiples of the critical temperature T_c.

given in table 6.1. Throughout the shaded area on the lower left-hand part of figure 6.1 the isotherms are all vertical and parallel to the pV axis. It is in this region of the diagram, and this region only, that we find a distinct phase change with the gas *condensing* to form a liquid. As the state moves down one of the vertical sections of an isotherm the volume decreases as gas condenses to form liquid and the pressure stays constant at the saturated vapour pressure for that temperature. It is evident that there exists a temperature called the *critical temperature* T_c below which all the isotherms pass through the region where condensation takes place and above which none of them do. The isotherm at the critical temperature has a single point of inflection where it becomes vertical and touches the shaded region. This point is called the *critical point*. The volume (of

Table 6.1 Boyle temperatures T_B/K, critical temperatures T_c/K, and the boiling temperatures T_b/K for various gases.

	He	Ne	Ar	H_2	N_2
T_B	22	134	410	106	323
T_c	5.2	45	151	33	126
T_b	4.2	27	87	20	77
T_B/T_c	4.2	3.0	2.7	3.2	2.6
T_c/T_b	1.2	1.6	1.7	1.6	1.6

one mole) and the pressure there are called the *critical volume* V_c and the *critical pressure* p_c. Collectively they are known as *critical constants*. If a gas above the critical temperature is increasingly compressed its properties will pass in a gradual and continuous way from those of gas to those of a liquid without any sudden phase change. Values of T_c for various gases are given in table 6.1. Figure 6.2 shows the quantity pV/qRT as a function of p for various gases at 0 °C. The approach to perfect behaviour at low pressures is demonstrated clearly as all the curves approach the value 1 in the limit $p \rightarrow 0$. Looking at these isotherms at pressures of a few atmospheres, where the $B'p$ term is the dominant one after the constant, we find the various gases showing different behaviour, some having positive and some having negative values of B'. Thus the fixed absolute temperature which is common to all the isotherms shown is, in this sense, a higher temperature for some gases than for others. We will return to this idea in connection with the law of corresponding states in §6.6.

Although the virial equation is very useful in relation to small departures from ideal gas behaviour, its mathematical form as a series expansion is not suitable for handling the more dramatic behaviour of real gases. For example, the infinite gradient of the pV isotherm at the critical point could only be represented by using an infinite number of terms, which is impractical.

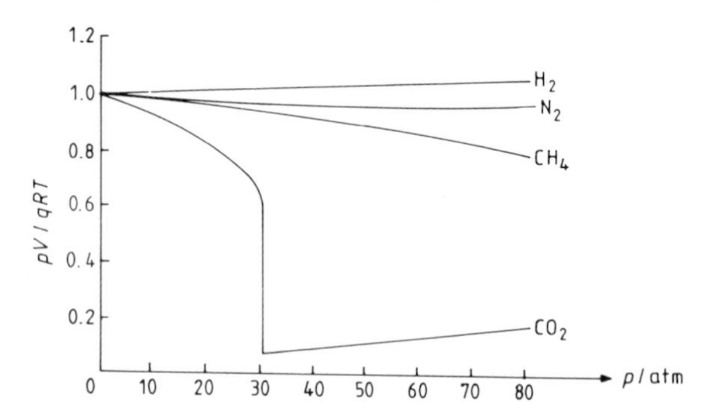

Figure 6.2 The variation of pV/qRT with p for various gases all at 0 °C.

6.2 Van der Waals' Equation of State

In 1873 the Dutch physicist van der Waals proposed a general equation of state for a gas, which for one mole has the form

$$\left(p + \frac{a}{V^2}\right)(V - b) = RT \tag{6.2}$$

where a and b are positive constants. To adapt the equation for q moles, V is replaced by V/q throughout on the LHS and the RHS is left unchanged as RT. This remarkably successful equation is still in general use. Two points are worth noting: (i) as shown in figure 6.3 it is able to describe the behaviour of gases over a surprisingly wide range of pressures and temperatures, and even gives an indication of the region of liquefaction, and (ii) the need to introduce the two constants a and b can be explained in terms of the volume of the molecules and the forces of attraction between them.

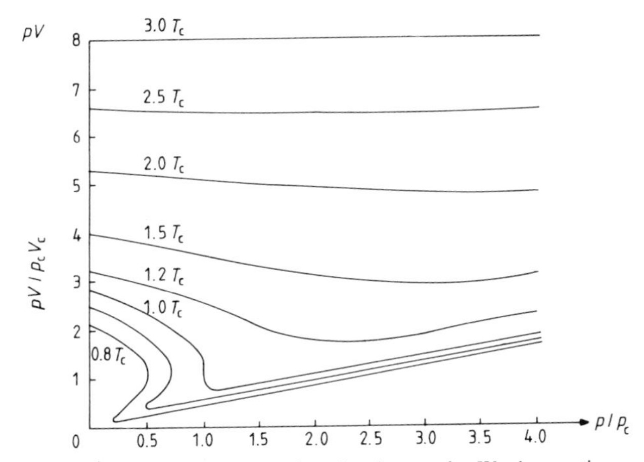

Figure 6.3 pV isotherms as given by the van der Waals equation.

Much of our remaining discussion of van der Waals' equation will concentrate on its connection with the second virial coefficient B'. At not too high pressures, we can use the relations $q^2a/V^2 < p$, and $qb < V$, to make a binomial expansion of the equation which, after some substitutions in the small terms using the approximate relation $pV = qRT$, leads to

$$pV = qRT\left[1 + \left(\frac{b}{RT} - \frac{a}{(RT)^2}\right)p + \ldots\right] \tag{6.3}$$

according to which the second virial coefficient is

$$B' = \left(\frac{b}{RT} - \frac{a}{(RT)^2}\right). \tag{6.4}$$

Given that a and b are both positive, it is already evident that the sign of B' will change with temperature. By setting B' equal to zero we find that the Boyle temperature is given by $T_B = a/Rb$. The van der Waals equation also produces one isotherm with a single point of inflection corresponding to a critical temperature T_c given by $T_c = 8a/27Rb$. Combining the two expressions leads to the relation $T_B/T_c = 27/8 = 3.375$. The values shown in table 6.1 are all within 20% of this result.

6.2.1 *Arguments for Volume and Pressure Corrections*

The volume correction b comes about because the finite volume of molecules reduces the space available for their motion in the vessel. Thinking of any molecule in particular, it is not possible for the centre of another to get closer than a distance of one diameter σ from the centre of the first. The volume from which it is excluded is thus $\frac{4}{3}\pi\sigma^3$ which is 8 times that of a molecule. The loss is shared between any pair of molecules which make a close encounter, so that finally the total 'lost' volume b in one mole of gas is given by

$$b = N_A \tfrac{2}{3}\pi\sigma^3. \tag{6.5}$$

Whilst it is not easy to be completely convinced by this particular argument (and better arguments can be made, such as the one in the next section) it is generally accepted that the existence of the b term is due to the volume of the molecules.

The implication is that the gas will behave as a perfect gas, provided that we use the corrected volume V_{per}, where $V_{per} = V + \delta V$, i.e. $V_{per} = V - b$. One can also enquire whether there is a pressure correction where $p_{per} = p + \delta p$. Here, p_{per} is what we would have calculated for a perfect gas using $\frac{1}{3}mn\overline{c^2}$. Van der Waals argued that there was such a correction due to the forces of attraction between molecules. As a molecule approaches the wall, it moves from a situation in the body of the gas where it has neighbours pulling on it from all directions, to one where, on reaching the surface, it only has neighbours on the side away from the wall which exert a net force pulling it back into the gas. The effect of this force is to slow the molecule down as it approaches the wall, causing it to deliver a smaller impulse when it arrives. The number of neighbours contributing to the retarding force is proportional to n, and the original momentum current per unit area of surface to which the correction is being applied is also proportional to n, leading to a pressure correction $\delta p \propto n^2$, i.e. $\delta p \propto (N_A/V)^2$. The sign clearly has to be such that $p < p_{per}$; hence we write $\delta p = +a/V^2$ so that a will be positive. Again this is a rather qualitative argument. A more detailed calculation will be given in §6.4. Finally, van der Waals' equation is obtained by substituting these expressions for p_{per} and V_{per} in the perfect gas equation $p_{per}V_{per} = RT$.

6.2.2 *Molecular Attraction to the Walls*

The attractive forces between the molecules and the walls do *not* influence either the pressure or the equation of state. All the time that a force of this kind is accelerating a molecule towards the wall, the wall is also being pulled towards the molecule and equal and opposite momentum transfers take place between them. When the molecule hits the wall the extra impulse delivered just cancels the momentum transferred during the run up. A similar cancellation takes place when molecules are ejected from the wall and move away into the gas.

6.3 Corresponding States

For some purposes it is useful to express p, V and T as multiples of the critical values p_c, V_c and T_c respectively by introducing a *reduced pressure* p_r, a *reduced volume* V_r and a *reduced temperature* T_r defined by the equations

$$p_r = p/p_c \qquad V_r = V/V_c \qquad T_r = T/T_c. \tag{6.6}$$

Experimental results show that isotherms for $p_r V_r$ as a function of p_r at various T_r are the same for most gases to within a few per cent. To this extent there is, after the scaling operation of (6.6), an approximate universal gas law for real gases in which they all depart from perfect gas behaviour in the same way and which is represented approximately by the isotherms of figure 6.1. It follows that if the two p_r values and the two V_r values are made equal for samples of two different gases, the samples will then have the same values for T_r. This is known as the *law of corresponding states.*

The van der Waals equation implies a law of corresponding states which is exact; first, it yields the expressions $p_c = a/27b^2$, $V_c = 3b$ and $T_c = 8a/27Rb$; then using these with the equations (6.6) to substitute back into van der Waals' equation leads to a *reduced equation of state*,

$$\left(p_r + \frac{3}{V_r^2}\right)(3V_r - 1) = 8T_r \tag{6.7}$$

which is the same for all gases. In fact, any equation of state with just two characteristic constants like a and b can be reduced to a universal equation in this way. In principle, if all the intermolecular force potential energy curves $V(R)$ could be reduced to exactly the same shape with two scaling factors, one for the V axis and one for the R axis, the law of corresponding states would in reality be exact; the fact that it is not indicates that the shapes of the curves for different molecules show a little more individuality.

As an example of the usefulness of the isotherms of figure 6.1, we will estimate the number of moles, q, in a gas cylinder with a volume of one litre containing methane at 100 atm and 20 °C. For methane the critical temperature is 191 K, hence $T_r = 293/191 = 1.53$. The critical pressure is 46 atm so that $p_r = 100/46 = 2.17$. Using the isotherm with $T_r = 1.5$, we see that on passing from $p_r = 0$ to $p_r = 2.17$ the value of $p_r V_r$ reduces by a factor of 0.86 and, since pV/qRT is always 1 at $p_r = 0$, it must have fallen to 0.86 at $p_r = 2.17$, giving $q = pV/0.86RT$. Inserting the given values on the RHS yields the answer $q = 4.8$ moles. If the perfect gas assumption had been made, we would have estimated q to be 4.2 moles.

It is interesting to note that kT_c is approximately equal to the quantity which is called the binding energy and shown as ΔE on the $V(R)$ curve in figure 2.1. It seems that when the kinetic energy of translation of the molecules is significantly larger than ΔE the molecules are always able to get away from each other and the gas cannot be made to condense, however much the pressure is increased.

6.4 Theory of *a* and *b* using the Maxwell–Boltzmann Distribution

In this section we will consider both a and b to be associated with pressure corrections. In Chapter 3 we established that the pressure of a *perfect* gas is given by the equation $p=\frac{1}{3}mn\overline{c^2}$. One way to avoid complications due to collisions between molecules in deriving this last expression is to use the method of §3.1, where we considered just those molecules which are about to hit the wall in the next small interval dt. They are then so close to the wall that the proportion of them which collide with another molecule before hitting the wall is negligible. However, to be consistent in this approach, we must acknowledge that the values of $\overline{c^2}$ and n to be used are those which apply at the extreme surface. In general, $\overline{c^2}$ is the same at *all* locations which are in equilibrium at the same temperature. At first sight this seems to be in conflict with the idea that the molecules slow down as they approach the wall, but this is the same paradoxical situation which was discussed when considering the isothermal atmosphere in §4.4. We can make use of the Maxwell–Boltzmann distribution (4.13) which shows that the number density n changes with position if the potential energy of the molecules changes. The argument of van der Waals has already implied that the potential energy at the surface is not the same as that in the bulk. Since these differences are confined to a relatively small fraction of the volume of the vessel near the surface, they have a negligible effect on the average number density n in the bulk of the gas, and we can write, for the number density n_s at the surface,

$$n_s/n=\exp(-\delta\text{PE}/kT) \tag{6.8}$$

where the RHS is the ratio of the Boltzmann factors for molecules at the surface and in the bulk, and δPE is the increase in potential energy of a molecule when it moves to the surface from the bulk (*excluding* the effects of attraction to the wall). The expression for the pressure is now $p=\frac{1}{3}mn_s\overline{c^2}$, which, on multiplying both sides by V, and using (6.8) to substitute for n_s in terms of n, leads to the result

$$pV=\tfrac{1}{3}mnV\overline{c^2}\exp(-\delta\text{PE}/kT)=qRT\exp(-\delta\text{PE}/kT). \tag{6.9}$$

From (6.9) we see that δPE is the key to calculating the virial equation.

6.4.1 *Potential Energy Due to the Excluded Volume*

Molecules of finite size have potential energy by virtue of being in a gas of moving molecules as opposed to being in a vacuum. This can be calculated from the work which needs to be done to bring a molecule into the gas, by taking the pressure times the volume, $\frac{4}{3}\pi\sigma^3$, of the sphere out of which the gas must be pushed to make room for the newcomer. To bring the molecule in at the extreme surface, we only have to push the gas out of one hemisphere of radius σ and the potential energy is only half of what it is in the bulk. Thus, δPE_1 due to the excluded volume is given by $\delta\text{PE}_1=-\frac{2}{3}\pi\sigma^3 p$. Inserting this in (6.9) and expanding the exponential as a power series, we obtain

$$pV=qRT\left(1+\frac{(2/3)\pi\sigma^3}{kT}p+\dots\right). \tag{6.10}$$

The coefficient of p on the RHS is a contribution to B' and can be written in the form $\frac{2}{3}\pi\sigma^3 N_A/RT$, which (being positive) may be identified with the first term on the RHS of (6.4), implying that $b=\frac{2}{3}\pi\sigma^3 N_A$, which is the same result as (6.5). Before leaving this calculation, it is worth noting that the finite size of the molecules increases the number density close to the wall, by a process in which molecules can be pictured as being pinned to the wall by the crowding of their neighbours. The effect has been related to a velocity-dependent potential energy which would be zero if all the neighbours were at rest exerting no pressure.

The increase in momentum current, due to the size of the molecules, has the same magnitude in the bulk of the gas, although the mechanism for the increase there is different, it being due to knock-on effects in collisions, where momentum is transferred forward almost instantaneously from the centre of one molecule to the centre of another.

6.4.2 *Potential Energy Due to the Attraction Between the Molecules*

We need to think in terms of the potential energy function $V(R)$ shown in figure 2.1. For any particular molecule, we are interested in the cases when a neighbouring molecule comes closer than the distance R_1 within which $V(R)$ is significantly different from zero. In the body of the gas, the probability that a neighbour lies in the distance range R to $R+\mathrm{d}R$ is $n4\pi R^2\,\mathrm{d}R$. Thus the average potential energy $\bar{V}$ of a molecule in the gas is given by

$$\bar{V}=-n\int_{\sigma}^{R_1}|V(R)|4\pi R^2\,\mathrm{d}R=-nI. \tag{6.11}$$

Equation (6.11) is also being used to define the integral I. For a molecule at the surface, the factor 4π is replaced by 2π because neighbours can only lie in one hemisphere, and the average potential energy is then only $\bar{V}/2$. It follows that $\delta\mathrm{PE}_2$ due to the attractive forces is $-\bar{V}/2$. Substituting this result for $\delta\mathrm{PE}$ in (6.9) and expanding the exponential gives

$$pV=qRT\left(1+\frac{\bar{V}}{2kT}+\ldots\right)=qRT\left(1-\frac{IN_A^2 p}{2(RT)^2}+\ldots\right). \tag{6.12}$$

The coefficient of p on the RHS is negative and can be identified with the second term on the RHS of (6.4) for B', which implies that $a=\frac{1}{2}N_A^2 I$.

Strictly, in (6.11) we should have allowed n to vary with the local potential energy $V(R)$ according to the Boltzmann factor as in (4.15). However, the changes introduced by doing this only influence the third and higher virial coefficients and the conclusion concerning the second is unchanged.

Specific Heat Capacities of Gases

7

7.1 Basic Principles

The specific heat capacity C of a substance under specified conditions is defined as $C=\mathrm{d}Q/\mathrm{d}T$, where $\mathrm{d}Q$ is the *net* quantity of heat needed to increase the temperature of unit mass of the substance by an amount $\mathrm{d}T$ under the given conditions. The first law of thermodynamics in the form

$$\mathrm{d}Q=\mathrm{d}U+\mathrm{d}W \tag{7.1}$$

tells us that the heat supplied has to provide any increase $\mathrm{d}U$ in the internal energy of the gas, and any work $\mathrm{d}W$ which the gas may have to do on the surroundings as it expands. Generally $\mathrm{d}W=p\,\mathrm{d}V$ where $\mathrm{d}V$ is the increase in volume. We define two specific heat capacities, the specific heat at constant volume C_V where no work needs to be done, and the specific heat at constant pressure C_p where there is work to be done:

$$C_V=\frac{\mathrm{d}Q_V}{\mathrm{d}T}=\left(\frac{\partial U}{\partial T}\right)_V \qquad \text{and} \qquad C_p=\frac{\mathrm{d}Q_p}{\mathrm{d}T}. \tag{7.2}$$

7.2 Internal Energy

The internal energy U of a gas is the total energy of all kinds except those due to external forces such as gravity, stored in a gas at equilibrium in a stationary vessel. In general, U may be expressed as

$$U=\sum_i \mathrm{KE}_i+\sum_i \varepsilon_i+\sum_i \bar{V}_i \tag{7.3}$$

where the summations are over all the molecules and where KE_i are their kinetic energies of translation, ε_i are the energies of excitation of the internal structures of individual molecules and $\bar{V}_i$ are the mean potential energies of the molecules due to forces exerted by their neighbours. As described in §2.6 a certain number of active degrees of freedom per molecule, f, can be associated with the internal energy, and in equilibrium at temperature T each molecule will have an average energy $kT/2$ for each of its active degrees of freedom. The internal energy can then be written as

$$U = NfkT/2 + \delta_{\text{int}} + \sum_i \bar{V}_i. \tag{7.4}$$

The translational motion will contribute three of the f active degrees of freedom and the remainder, if any, will come from the internal excitations. The term δ_{int} is a small term representing those internal excitations for which the degrees of freedom are inactive or at least largely suppressed. The $\bar{V}_i$ are also small and associated with degrees of freedom which are strongly suppressed.

7.3 Specific Heat Capacities of Perfect Gases

7.3.1 *Joule's Law*

The $\bar{V}_i$ are evaluated according to (6.11) and are proportional to the probability of finding neighbouring molecules close to the one of interest. They are the only terms contributing to the internal energy which depend on the volume V of the vessel. In the low density limit of a perfect gas where the volume is large, the molecules are so seldom close together that the $\bar{V}_i$ become negligible. Thus, the internal energy U of a fixed mass of a perfect gas is given by just the first two terms of (7.4), and it is independent of the volume, in agreement with Joule's law.

Hence, when a perfect gas is heated the change in U is the same if the heating is at constant pressure where V increases, or at constant volume, i.e.

$$\left(\frac{\partial U}{\partial T}\right)_p = \left(\frac{\partial U}{\partial T}\right)_V \qquad \text{for a perfect gas.} \tag{7.5}$$

7.3.2 *Heat Capacities at Constant Volume and Constant Pressure*

From (7.2), (7.4) and (7.5), omitting the last term of (7.4) for a perfect gas and also the other small term, we find that C_V is given by

$$C_V = \frac{\mathrm{d}Q_V}{\mathrm{d}T} = \left(\frac{\partial U}{\partial T}\right)_V = \tfrac{1}{2}Nfk = q\tfrac{1}{2}fR. \tag{7.6}$$

It follows that C_V (per mol) $= \tfrac{1}{2}fR$, C_V (per kg) $= \tfrac{1}{2}f \times 10^3 R/M_{\mathrm{M}}$, and that C_V (per molecule) $= \tfrac{1}{2}fk$.

When heated at constant pressure a gas must expand. To calculate the work done, $p\,\mathrm{d}V$, we differentiate the perfect gas equation $pV = qRT$ and obtain

$$\mathrm{d}p\,V + p\,\mathrm{d}V = qR\,\mathrm{d}T. \tag{7.7}$$

At constant pressure this becomes $p\,\mathrm{d}V = qR\,\mathrm{d}T$ so we substitute $qR\,\mathrm{d}T$ for $\mathrm{d}W$ in (7.1) giving $\mathrm{d}Q = \mathrm{d}U + qR\,\mathrm{d}T$. Finally, on dividing both sides by $\mathrm{d}T$ we have

$$C_p = \frac{\mathrm{d}Q_p}{\mathrm{d}T} = \left(\frac{\partial U}{\partial T}\right)_p + qR. \tag{7.8}$$

The first term on the RHS of (7.8) is seen from (7.2) and (7.5) to be C_V, hence

$$C_p = C_V + qR = q\tfrac{1}{2}fR + qR = q\tfrac{1}{2}(f+2)R. \tag{7.9}$$

From (7.6) and (7.9) the ratio γ of the specific heats is

$$\gamma = \frac{C_p}{C_V} = \left(1 + \frac{2}{f}\right). \tag{7.10}$$

7.3.3 *Adiabatic Equation for a Perfect Gas*

If a gas expands adiabatically, i.e. under conditions where no heat can flow into or out of the sample, then $dQ=0$ and (7.1) shows that $dU = -dW$, which may alternatively be expressed as $C_V\,dT = -p\,dV$, meaning that the work needed for the expansion must come from the internal energy. Combining this last result with (7.7) and (7.9) to eliminate dT and qR leads to the equation $p\,dV = -\gamma V\,dp$ which integrates to give the *adiabatic equation for perfect gas*

$$pV^{\gamma} = \text{constant}. \tag{7.11}$$

(When we are interested in an *isothermal expansion*, which is one that takes place while the gas is maintained at constant temperature, the appropriate equation to use is Boyle's law.)

7.3.4 *Monatomic Perfect Gases*

The rare gases are monatomic and at densities of the order of that of air or less they behave as perfect gases to a precision of better than 1 part in 10^3. There are no significant internal excitations at normal temperatures and the internal energy U comprises just the kinetic energies of translation of the atoms. Thus $f=3$, $C_V = 3R/2$ per mole, $C_p = 5R/2$ per mole, and $\gamma = 5/3$. When such a gas is heated at constant pressure 60% of the heat supplied goes into increasing the internal energy and 40% goes into work on the surroundings.

7.3.5 *Diatomic and Polyatomic Gases—Molecular Rotations*

For all molecules other than monatomic ones, excitations of the molecule make an important contribution to the internal energy and the specific heat capacity. At normal temperatures the relevant excitations are restricted to rotations of the molecules about their centres of mass, which we will consider first, and internal vibrations involving the relative positions of the component atoms.

The kinetic energy of rotation of a 3-dimensional body may be expressed as $E_R = \tfrac{1}{2}I_x\omega_x^2 + \tfrac{1}{2}I_y\omega_y^2 + \tfrac{1}{2}I_z\omega_z^2$, where ω_x etc are the angular rates of rotation and I_x etc are the moments of inertia about the axes x, y and z through the centre of mass. The quantised energy states associated with rotation about the x axis, for example, have spacings of the order of magnitude of $\hbar^2/2I_x$ where $\hbar$ is Planck's constant over 2π. The three independent squared terms of E_R indicate three degrees of freedom for the rotation of a molecule. However, linear molecules, which include all diatomic molecules, are a special case in which the degree associated with rotation about the axis of the molecule (which we have chosen to be the z axis in figure 7.1) is never active at normal temperatures; such a rotation would amount to electronic or nuclear excitation, and the spacing between the quantum states is then much too large compared with kT. Considering molecular rotations about axes which are not special in this way, diatomic hydrogen has the smallest I value and the largest energy spacing, and for kT to be comparable in magnitude to it, so that the degrees of freedom are becoming active, $T \simeq 100$ K. In all other molecules the spacings are smaller and the corresponding temperatures even lower. Thus, at room temperature all rotations are active except those about the special axis of linear molecules. The latter, therefore, have two active rotational degrees while nonlinear molecules have three and monatomics have none.

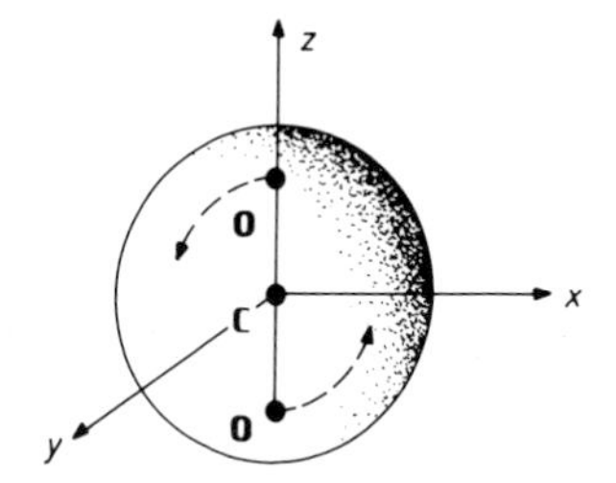

Figure 7.1 A linear molecule CO_2 with the molecular axis along z. Rotation about the y axis would involve the oxygen atoms moving round the broken circle.

7.3.6 *Diatomic and Polyatomic Gases—Molecular Vibrations*

The simplest example of molecular vibration is provided by diatomic molecules. They have just one *mode* of vibration in which the two atoms move with simple harmonic motion along the molecular axis as though they were connected by a light spring. Any mode involves two degrees of freedom, one for the kinetic energy and one for the potential energy. The quantised contribution to the energy of the molecule is $E_n = (n + \frac{1}{2})h\nu$ where ν is the vibration frequency, n is a quantum number which is an integer, and h is Planck's constant. The energy level spacings are seen to be $h\nu$. Again H_2 has the largest spacing with $h\nu = 8 \times 10^{-20}$ J. The temperature at which the degrees of freedom become active is where $2kT \simeq h\nu$, which for H_2 is about 3000 K. For N_2 and O_2 the change comes at 1000 K, and in a heavy molecule like I_2 the change is at about 150 K. Thus, in air at room temperature the vibrational degrees are not active; on the other hand, there are two active rotational degrees and we conclude that $f = 5$, $C_V = q\frac{5}{2}R$, $C_p = q\frac{7}{2}R$ and $\gamma = \frac{7}{5} = 1.4$. The specific heat capacities of hydrogen and nitrogen over a range of temperatures are shown in figure 7.2.

For more complicated molecules the first problem is to find out how many modes of vibration there are. This can be done by arguing that the total number of degrees of freedom per molecule associated with *translational kinetic energy of the component atoms* is always $3N_m$ where N_m is the number of atoms in the molecule. A number j of these is accounted for by the translation of the whole molecule (three) and by its rotations (two or three depending on whether the

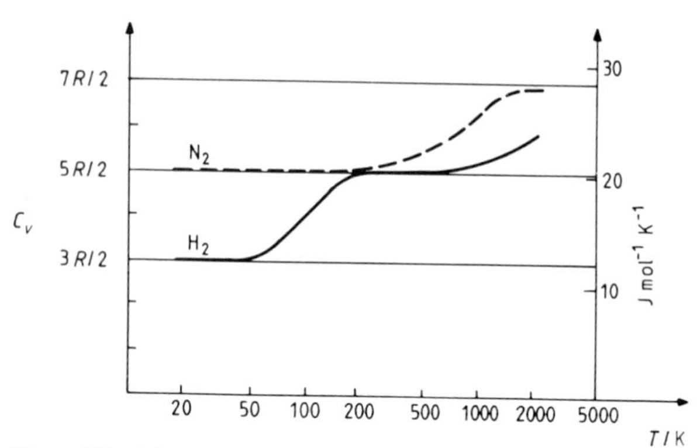

Figure 7.2 The specific heat capacities of hydrogen and nitrogen as a function of temperature.

molecule is linear or not). Thus, for linear molecules $j=5$, and for nonlinear molecules $j=6$. Each vibrational mode accounts for one of the remaining $3N_m-j$ degrees, hence the number of modes must be $3N_m-j$. Whether the associated degrees are active at a given temperature can only be decided when the frequencies of the modes are known.

7.4 Heat Capacities at Higher Pressures and Densities

As usual, the differences compared with the perfect gas case are due to the intermolecular forces and collisions. Consider the relation (7.9) $C_p-C_V=qR$ for q moles of perfect gas. At higher pressures where gases are no longer perfect we can approximate to the equation of state with $pV=qRT(1+B'p)$. The quantity C_p-C_V is $\mathrm{d}W_p/\mathrm{d}T$ which can be found by differentiating the last expression for pV to give $p(\partial V/\partial T)_p=qR[1+B'p+T(\mathrm{d}B'/\mathrm{d}T)p]$ which replaces what is just qR in the case of a perfect gas. For N_2 at room temperature and 1 atm, the fractional change is only of the order of 10^{-3}, whereas at 100 atm the change is tens of per cent. In §7.2 we have already seen that intermolecular forces contribute the quantity $\sum_i \bar{V}_i$ to the internal energy. Again, this extra term, which is negligible for a perfect gas, becomes significant at higher pressures. It influences the value of C_V through the first of the relations (7.2), and through (7.3).

7.5 Joule–Kelvin Expansion

In his early experiments on the energy content of gases Joule was not able to detect any change with variation of pressure and volume at a fixed temperature. This led him to Joule's law, which we now know to be true only in the case of a perfect gas. Subsequently, he used a more sensitive method to search for such changes suggested by Lord Kelvin. The apparatus is shown in figure 7.3. The

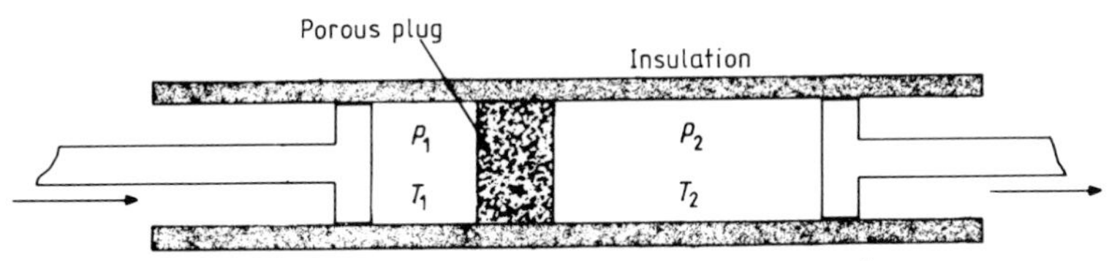

Figure 7.3 Apparatus for a Joule–Kelvin expansion.

pistons are controlled so as to keep p_1 and p_2 constant with $p_1 > p_2$. Initially, all the gas is on the left-hand side of the porous plug in the state p_1, V_1, T_1. Then, due to the pressure difference, it flows steadily through the porous plug until finally it is all on the right in the state p_2, V_2, T_2. In general, T_2 is found to be different from T_1. The total work done on the gas by the piston on the left is $p_1 V_1$ and the total work extracted by the piston on the right is $p_2 V_2$. The net work done by the gas against external forces is $p_2 V_2 - p_1 V_1$. Insulation round the apparatus ensures that no heat flows into or out of the gas.

We will now make a simple estimate of the energy changes in a Joule–Kelvin expansion using the virial equation of state with the van der Waals expression (6.4) for the second virial coefficient and assuming the pressures are not so high that we need to use any higher coefficients. The work done against external forces is then $p_2 V_2 - p_1 V_1 = RTB'(p_2 - p_1) = (b - a/RT)(p_2 - p_1)$. We also need to consider work done internally which, in an expansion, causes energy to be taken from the kinetic energy of translation and from the internal excitations of the molecules, and given to the potential energy of intermolecular attraction. Although this redistribution leaves the total internal energy unchanged, it slows down the molecules and must therefore lower the temperature. To estimate this increase of potential energy we calculate the work done against the van der Waals pressure correction δp, where for one mole $\delta p = a/V^2$. This internal work done is then

$$\int_{V_1}^{V_2} \frac{a}{V^2}\,\mathrm{d}V = \frac{a}{V_1} - \frac{a}{V_2} \simeq \left(\frac{a}{RT}\right)(p_1 - p_2).$$

Adding together the internal and external work we find that the total energy removed from translation and excitation of the molecules when one mole passes through a Joule–Kelvin expansion may be expressed as

$$\left(\frac{2a}{RT} - b\right)(p_1 - p_2). \tag{7.12}$$

The temperature will rise or fall depending on whether this quantity is negative or positive. It is zero at the *inversion temperature* $T_i = 2a/Rb$, which is just twice the Boyle temperature given by the same model.

Not long after the early experiments it was realised that the Joule–Kelvin expansion provided an extremely important technique for refrigeration where temperatures in the range 4–250 K have to be produced as in the liquefaction of

O_2, N_2, H_2 and the rare gases. Essentially, a gas can be used to transfer heat in a way that resembles the use of a sponge to transfer water from a floor to a sink. When a gas which is below its inversion temperature expands through a nozzle it cools and can then soak up some heat; subsequently it may be piped away to a place where it is compressed in thermal contact with a sink which receives the heat expelled.

In a *Joule expansion* or *free expansion* which had been used by Joule in earlier experiments, a gas is allowed to increase its volume by rushing into a vacuum space which is thermally insulated. Under these conditions no external work is done and no heat flows into or out of the gas. It follows that the total internal energy U of a gas remains unchanged in a Joule expansion. Except in the perfect gas limit, the gas always cools, however, as internal energy is redistributed by transfer from molecular translation and internal excitation to the intermolecular potential energy. The amount transferred for one mole is the internal work, estimated above to be $(p_1 - p_2)a/RT$ per mole.

7.6 Speed of Sound

The speed of sound c_s in an elastic medium is given by the relation

$$c_s = \sqrt{\frac{\text{elastic modulus}}{\text{density}}}. \tag{7.13}$$

In a gas the stress is dp and the strain is $-dV/V$ and, as usual, the elastic modulus is the ratio of the two. An analysis of heat flow in a sound wave shows that the compressions and rarefactions are adiabatic. We will restrict ourselves to a perfect gas in the remainder of this section. By differentiating (7.11) we find that the adiabatic bulk modulus is γp. (The isothermal bulk modulus found by differentiating Boyle's law is just p.) Substituting into (7.13) and using $\rho = qM_A/V$, we obtain

$$c_s = \sqrt{\frac{\gamma p}{\rho}} = \sqrt{\frac{\gamma p V}{q M_A}} = \sqrt{\frac{\gamma RT}{M_A}}. \tag{7.14}$$

Comparing this with equation (3.24) shows that the ratio of the speed of sound to the RMS speed of the molecules is equal to $\sqrt{(\gamma/3)}$, which, for perfect gases, is in the range 0.57–0.75. Given that the pressure waves must be transmitted by the molecules colliding with each other, it is reasonable that the speed of sound cannot be greater than that of the speed of the molecules. It is interesting that the process is so efficient that the two speeds are very similar.

Index